SUGAR

SUGAR
The Grass that Changed the World

Sanjida O'Connell

First published in Great Britain in 2004 by
Virgin Books Ltd
Thames Wharf Studios
Rainville Road
London
W6 9HA

A catalogue record for this book is available from the British Library.

ISBN 1 85227 034 9

Typeset by Phoenix Photosetting, Chatham, Kent
Printed and bound in Great Britain by Creative Print and Design

CONTENTS

AUTHOR'S NOTE AND ACKNOWLEDGEMENTS

Lozenges of coloured light from the stained-glass windows fell across the table. A monk glided over to us, folded his hands together and passed us a menu. Vaughn scrutinised it through his rose-tinted glasses, sighed, chose and then, after I had ordered, launched into a diatribe against sugar.

I was making a documentary on the science of kissing; we were in Atlanta filming a demonstration of thirty different sorts of kisses and Vaughn Bryant, an anthropologist from Texas A & M University, had flown over for the day so that I could interview him – and so he could watch the kissing demo. He'd spoken eloquently on the subject of kissing: he believed kissing was a learned phenomenon and that many tribes only kissed because they'd seen Westerners do so. He told me that Roman men were permitted to kiss women in the street because it was forbidden for women to drink and by kissing them they could check for alcohol on their breath. He said the Celts believed kissing under the mistletoe was the equivalent of a legally binding promise to marry the woman you had kissed, and that you'd be sued if you didn't honour your promise, but I thought this was an example of the ubiquity of American litigiousness clouding our understanding of an ancient Celtic custom.

To thank him I'd brought him to a restaurant set in a church where the waiters were dressed as monks and the food was tastefully presented. Vaughn's real job is as a palynologist, a person who studies pollen – in his case, fossilised pollen. He told me that his problem with sugar began about thirty years ago. He had grown overweight and was self-conscious about it. He was about to begin an archaeological dig in Texas when an idea struck him. He shared it with his students – while they were excavating the remains of cave people, why not eat as they had? The students – I'm not sure how readily – agreed. Vaughn had a pretty good idea what humans had eaten in that part of the world because he analyses coprolites – 10,000-year-old fossil faeces. So in the mornings he and his students did not eat breakfast, but instead foraged for berries as they walked to the dig. They ate cacti, roots, grubs, lizards and rats. Unsurprisingly, Vaughn lost some weight.

Back at campus he wondered whether he could continue with the diet. Obviously, roast rat with a side of steamed cactus was not going to work in a university setting, so he thought about the type of food his ancestors had been eating. He came to the conclusion that they ate a lot of fruits and vegetables, lean meat occasionally and the odd egg, but no refined carbohydrates, no dairy products, little fat and certainly no sugar. Trying to translate this into the kind of food you can buy at the supermarket was initially a challenge. He was shocked to discover that almost all processed food has sugar in it. Even baby food contains sugar: anyone of a slightly cynical disposition might think this was a ploy on the part of manufacturers to hook consumers at as early an age as possible. Sugar, after all, is not required by babies; it is not even a nutritional necessity for adults.

Vaughn told me of an experiment reported in William Dufty's book *Sugar Blues*, in which half the mental patients in a hospital were deprived of sugar and half were allowed to eat as many sweets as possible. After two years the patients who had eaten the sugar had lower IQs, their memories were impaired and their mental health had deteriorated in comparison with the other patients. He said sugar was responsible for obesity, heart problems, diabetes, hyperactivity and schizophrenia. It made sense: our bodies are not designed for sugar. We have evolved to subsist on meagre rations – eating highly processed and refined foods means our bodies think we have come across an unbelievable bonanza and store the energy for leaner times. We have not yet found a way to let our bodies know that, for many of us living in developed countries, the bonanza is all around us, every day. Still on the caveman diet, Vaughn is fit and trim despite the fact that he's edging up to retirement. Neither of us had dessert.

Back home, intrigued by his story, I looked sugar up on the Internet. I was amazed to discover that sugar is a grass. It looks like bamboo and grows several metres in height. How, I thought, have we gone from being cave peoples scrabbling around for roots, nuts and berries, plus the odd lizard, to cultivating a type of grass that we grind down in order to extract its juice and boil until we have fine, white crystals and then eat this strange food stuff at practically every meal? How did this type of grass end up being responsible for the deaths of millions of indigenous people in the countries to which sugar cane was transported? How did whole nations of Africans lose their lives in the growing of it? Why did entire European economies wax and wane according to the harvesting of this crop, which was entirely dependent on a lot of rain, free labour, low taxes and the absence of plagues of caterpillars? And how has it come to pass that we cannot survive without this sweet concoction and, as a result, are suffering from chronic obesity, a situation that is costing the health budgets of nations millions worldwide?

It is thus to Vaughn Bryant that I owe the inspiration for this book. I have another major debt that I must also acknowledge. At about the same time as I started working on the kissing programme, I began a relationship with James Fair, an editor at *BBC Wildlife* magazine. By the time I had researched my idea for a book on sugar and agreed with Virgin Books to publish it, James and I were married. I now, somewhat ironically, found myself wedded to a sugar family. James's father, Neil Fair, his maternal grandfather, Kenneth Brown, his uncle, Peter Brown, and some of his other relatives all worked for Tate & Lyle. It is due to Jennifer and Neil Fair's help and generosity – in spite of their amusement over my chosen subject matter – that I have been able to meet Michael Grier, Community Relations Manager for Tate & Lyle, who kindly let me use the company library, and ex-directors Michael Attfield, Colin Lyle and Sir Saxon Tate. These men and their wives were good enough to welcome me into their homes and answer my simplistic questions; Michael Attfield, in particular, was probably shocked to see such a profound display of ignorance over the futures market. Colin Lyle and Saxon Tate put me in touch with Chuck Vlitos, ex-director of Tate & Lyle's research laboratories at Reading. Neil and Peter both lent me books, and Peter provided me with useful contacts in Barbados. Their views are, of course, their own and should not be taken to represent the views of Tate & Lyle.

I would also like to thank all those academics who gave their time and expertise and sent me papers and references. In particular, I would like to thank the following: Sandra Bellamy, Gordon Birch, Harm de Boer, Angélique D'Hont, Jock Galloway, Matt Griffiths, Peter Havel, Patrick Holford, Anthony Kennedy, Frieder Lichtenthaler, Mike Lindley, Seshagiri Rao, Aubrey Sheiham and Toni Steer. Louise Osborne thoroughly, meticulously and swiftly carried out some preliminary research for my original proposal. I have used many books and papers, but the most useful by far on the history of sugar has been Noël Deerr's two volumes on sugar published in 1949 and 1950 – his breadth of knowledge and attention to detail is unparalleled. Philippe Chalmin's Master's thesis on Tate & Lyle proved immensely helpful, as it is dense with a richness of detail unrivalled by any other books on the subject. Bristol University's excellent library held many key books and Bristol Central Library's staff have been extremely efficient, procuring those the university did not have. Kew Gardens kindly let me visit their Economic Botany library. I'm very grateful to the staff of the Christopher Columbus Museum in Porto Santo who opened it although it was closed. In addition, David Smith and Annette

Green, my agents, have provided me with unflagging support and dedication, as have my family and newish husband! James Fair, my father, James O'Connell, and my sister Sheila have all read parts of my book and made useful comments. I would also like to thank Kirstie Addis for her thoughtful editing and Ian Allen for his careful copyediting, as well as the rest of the team at Virgin Books. Any mistakes are my own.

Note: All monetary figures are given as the value they had at the time, unless otherwise stated.

INTRODUCTION

It was thought singularly impolite not to wear a hat. In fact, the whole uniform, pith helmet and white suit included, was a requisite. I was watching a dying breed – the last great white sugar farmers from a bygone era – on a flickering reel of black-and-white film in the stables of a Jacobean mansion on Barbados. The film was taken in 1935 by plantation owner Laurence Cave but only discovered 45 years later by his grandson. These ten precious reels documented the final days of nonautomated sugar production. They depicted a windmill made of coral with canvas sails that small Barbadian boys climbed aboard for a hair-raising home-grown roller-coaster ride as the internal wooden cogs crushed sticks of sugar cane to extract its molasses-dark syrup; guinea fowl wandered in front of the mill, and horses toiled through the fields, pulling carts laden with freshly harvested sugar cane; sugar-cane juice bubbled in *taches* – huge, round copper pans almost as high as a person.

I had come to Barbados to research the story of sugar cane from its origins ten thousand years ago to the present day and, as I sat and watched this amateur film, I realised that Barbados *was* the history of sugar cane, in microcosm. I had driven through a tunnel created by the heavy branches of age-old mahogany trees to reach St Nicholas Abbey, a Jacobean mansion, complete with an English knot garden crammed with herbs and overblown roses, and fireplaces and chimneys, requested by an owner as yet unaccustomed to the heat of the Caribbean. The house had been built in 1650, less than 25 years after the English had seized Barbados and around the time they had decided that sugar was to be their main cash crop. Sugar cane, however, had probably been on the island long before it was reintroduced. It had probably been brought by tribal peoples, just as they had originally transported it around the world at the dawn of its domestication. On Barbados the people who brought it were the original inhabitants, the Amerindians, who were wiped out by the Spanish before the English appropriated the island.

Until recent technological innovations, sugar cane was an incredibly labour-intensive crop that needed to be planted, weeded, manured, irrigated and harvested by hand, before it was crushed for its juice, which was then boiled to extract crystals of sugar. In order to make this economical, the English, like other colonial powers, sent African slaves to Barbados. The majority of Barbadians alive today are descended from these original slaves and some of the remaining white plantation owners are related to the first white settlers. St Nicholas Abbey, like many other wealthy plantations, had its own windmill for crushing the sugar cane and its own factory for boiling the sugar-cane juice to extract unrefined muscovado sugar. The remnants of these buildings are still here today. The house itself feels dark, hemmed in as it is by the giant mahoganies, grown to house and furnish the descendants of the plantation; but also because the house, the factory and windmill still exist today in a charmed circle, surrounded as they are by 220 hectares of sugar cane, a grass that can reach heights of six metres, with a jointed bamboo-like stem and thick tassels of cream-coloured flowers.

Today the island itself is still covered by vast fields of sugar cane and the history of the island is laced with the story of black slaves, white plantation owners and sugar. In the Barbadian Museum is an appeal for a runaway slave, which was published in the *Barbados Mercury* on Saturday 12 April 1788. It describes her as 'silly looking' with 'slim legs and a round belly'. Later, in 1834, the slaves were emancipated in the British Empire; in 1838 they made up the following song ('Jin-Jin' means Queen Victoria, 'licks' are beatings and 'lock-up' is a jailing):

Lick and lock-up done wid,
Hurrah fuh Jin-Jin;
Lick and lock-up done wid,
Hurrah fuh Jin-Jin.

God bless de Queen fuh set we free,
Hurrah fuh Jin-Jin;
Now lick and lock-up done wid,
Hurrah fuh Jin-Jin.

In spite of, or perhaps because of, the horrors of slavery, places like the Caribbean and the southern states of America now have a multi-cultural society; without sugar, it could be argued, we would not have the blues or rock and roll, musical genres that had their birth in the slave-owning counties of the USA. It is also no surprise that Barbadians, surrounded by sugar, have one of the highest rates of diabetes in the world. When I had a traditional meal, which included sweet potato pie made with added sugar, followed by a four-tiered sponge cake with butter cream and fondant icing, I could see why.

Without sugar, whole areas of land, be it in Barbados, Thailand or Australia, wouldn't have been altered so dramatically as native forests were felled to make way for field upon field of sugar, with the concomitant rise in pollution from the pesticides, herbicides and fertiliser this hungry crop requires. Today Barbados, a coral atoll in the Caribbean, still has its cap of sugar cane but, like many small countries facing pressure from imminent changes to EU legislation, sugar cane soon will no longer be economical to grow and export to its traditional sweet-toothed importer – Britain.

'People complain to me the whole time. They feel demoralised and demotivated. The sugar industry is running at a loss,' Harm de Boer tells me. A lean man whose ginger hair is greying, originally from Holland, he speaks English with a Dutch accent and a Scottish twang. A trained agronomist, he arrived in Barbados in 1981 to discover that the legacy of slavery still hung over the sugar-cane industry. Some Barbadians are still racist, referring to their workers as 'niggers', he adds. As soon as Barbados became independent in 1868, yields plummeted and have never recovered. Now de Boer, thinking of the imminent legislation and falling yields, has other reasons to be concerned.

De Boer's office is the nicest he's ever had. It's on the top floor of an old plantation house, and has a wooden floor, cream walls and cool jalousies – small roofs over each window so they can be kept open even in a storm without the rain running in. There are windows on three sides overlooking acres of sugar cane. But these are no ordinary fields of cane: de Boer works in the Agronomy Research and Variety Testing Unit and stretched below him are 3,500 experimental strips of sugar cane. An unusual mixture of quiet precision and enthusiasm, he's hoping that these plants will rescue Barbados.

Once sugar cane was a wild grass that contained very little sugar at all. Over many generations, people bred it to become sweeter and sweeter and learned to extract crystals of pure white sugar from its sap. While de Boer's colleague Dr Anthony Kennedy is scientifically continuing this process to create the world's sweetest sugar cane, de Boer is hoping that sugar cane will also yield other products, such as lignin, a chemical valued by pharmaceutical companies, or fibre to fuel power stations or make paper and cardboard.

People once desired sugar so badly that whole nations were wiped out, countries went to war, cultures were destroyed, trade blossomed and multinationals bloomed. Now there are surplus sugar mountains and sugar lakes; people in the West have grown obese through sugar indulgence. It is a quiet irony that the very plant we have manipulated for centuries to provide us with the maximum amount of sugar could well now be used to generate electricity and shore up our medicine cabinets.

As I watch the island of Barbados spinning out of view from the plane window, with its idyllic beaches, fields of sugar cane, jagged limestone outcrops and houses the colour of boiled sweets, modelled directly on the chattel houses the slaves once dwelled in, I realise how much sugar cane has changed us. More than any other crop, be it cotton or cocaine, sugar has shaped our culture, landscape, politics, geography, economics, race, music, health, the very food we eat and what we drink in a way that no other commodity has throughout human history. This is the story of sugar.

1. STONE HONEY

The juice of the sugar-cane, if the stalk is chewed with the aid of the teeth, increases the semen, is cool, purges the intestines, is oily, promotes nutrition and corpulency and excites the phlegm.
Charaka, in *Charaka-Samhita*, AD 78

God must have a sweet tooth. Sugar cane has been transported, almost without fail, throughout the world by men and women carrying faith and this overgrown grass. The handmen and maidens of the Gods belonged to practically every major religion – Hinduism, Buddhism, Christianity, Islam – and, wherever they took their message, they carried sweetness.

It all began on New Guinea ten thousand years ago on a small island in the South Pacific. The story goes that, once upon a time, two fishermen, To-Kabwana and To-Karvuvu, found a piece of sugar cane in their net. They threw it into the sea, but as they hauled their net in on the second day they saw that the same section of sugar cane was tangled in the folds. They cast it into the water, but, at the end of their third day of fishing when they pulled their net in, the cane was again snarled at the bottom. Thinking that this must be an omen, they kept it and planted it. From the stubby knob of cane grew a thick, strong plant. A few weeks later a beautiful young woman stepped from between the dense cluster of leaves. She cooked a meal for the men that night but as the moon rose she slipped back between the folds of cane. This became routine until one day To-Kabwana caught her before she could retreat into the sugar cane. He asked her to become his wife and she consented. They had children together and these children became the founding members of the whole of the human race.

Or at least, this is what the Polynesians believe. And *to* is their word for sugar cane. The Polynesians have a number of legends like this in which sugar cane and the genesis of humanity are intertwined; they show how important sugar cane is to the people from the lands where sugar cane has its origins.

Sugar cane is a giant member of the *Gramineae*, the grass family, often growing up to six metres tall. Its closest living relatives are maize, rice and sorghum, but sugar cane looks like bamboo: a long, thick stem joined in segments like an insect's leg. Some are thick burgundy with the midnight-grey glaucous bloom of a plum, or vermilion, with turquoise joints; others are flushed pink, with a pearly suffusion at the end of each segment, the colour of a crustacean's inner claw. Yet other varieties can be grasshopper-green or mustard-yellow, tinged with the brown of dying leaves. One kind, *badila*, is almost blue-black, while another has a glassy, white transparency. Wild sugar cane, *Saccharum spontaneum*, looks more like grass: it is tall, slender and vigorous and its sugar content is low.

But the grass that changed the world was Creole. The Creole cane is a freak of nature – it's sterile; it cannot survive without human help; it's pathetically susceptible to disease and, above all, its thick, segmented stem is packed full of sugar. All plants obtain their energy from the sun in a process called photosynthesis. The result is a simple sugar, glucose, which the plant uses to make its cells. To tide them through winter, for instance, many plants store their energy by converting glucose to starch and secreting it in special organs – potatoes are literally balls of starch. But very few plants store glucose as sucrose, a more complex sugar, which is the white sugar you buy in packets. Sucrose is an odd choice as a means of storing excess energy because it has to be dissolved and held in liquid form.

By a quirk of nature, some ancestor of sugar cane mutated and stored some of its energy as sucrose in the juice of its stem, instead of as starch in its roots. Nevertheless, that sucrose was used after the plant had flowered to help it grow the following year. But then, more than 8,000 years ago, someone tasted that cane and noticed that it was sweet. Instead of allowing the cane to continue to use up its reserves of sucrose, they cut it down and chewed on its sweet length. They kept it and grew it in their gardens, each year selecting the sweetest canes for chewing. This continued for thousands of years and, since the cane was unable therefore to use up its reserves, it became sweeter and sweeter until a fifth of its stem contained pure sucrose.

One other plant, apart from sugar cane, also stores sucrose in sufficient quantities to be used commercially – the sugar beet. Refining and growing sugar beet are complex, costly operations, but sugar produced from beets would eventually challenge the supremacy of sugar from sugar cane because it can be grown, harvested and refined in Europe. Sugar cane can only be grown in the tropics; the refineries, which are expensive to build and run, are almost always in developed nations.

Ancient peoples learned to extract sugar from sugar cane by crushing the plants' stems to procure its syrupy juice, at first by using primitive pestle and mortars; later wooden and stone mills were invented that were powered by animals, people or the wind. This juice is the colour of swamp mud, and tastes of liquid brown sugar, heavy with caramel and molasses and with a harsh, green, saplike aftertaste. This liquid would be boiled and the water allowed to evaporate until dark-brown sweet crystals were left.

Throughout the history of sugar there have been technological advances, such as the one in Barbados, called the Jamaica train. This was a series of copper pans; syrup was ladled from one vessel to another as more and more water was evaporated. Reagents were added, such as charcoal or bull's blood, to precipitate out the impurities; the final supersaturated syrup was left to dry in clay vessels that allowed the sugar to crystallise out. Even today most sugar-producing countries refine sugar to this extent. Now, of course, the technology is much improved and vacuum pans are used. The syrup is heated with steam at high pressure in giant stainless-steel vessels in order to evaporate the water at a lower temperature. The end result is a pale-brown sugar that is around 97–99 per cent pure.

This sugar is then sent to refineries, such as Tate & Lyle's in London, for those final impurities to be removed. The process is similar – the sugar is melted in water; the water is evaporated and the sugar is crystallised out and dried, but, at each stage, another fraction of a percentage of impurities are removed. These refineries produce the purest food substance known – almost 100 per cent white sucrose. It's a complicated process with an underlying simplicity: as Mark Twain said in *Life on the Mississippi* on 'The process of making sugar – the thing looks simple and easy.'

There has never been anything like it before – a natural substance produced by a plant that, using rudimentary technology, can give a fast shot of the substance most desired and craved by anyone anywhere – for even a newborn baby loves sugar. Yet our hominid ancestors would never have experienced sugar – ripe fruit and occasional honey was the closest early *Homo sapiens* would have come to this mind-blowing sweetness.

Leaving myths to one side, one of the reasons why people suspect that sugar cane originated in New Guinea is largely due to the relationship a species of beetle has with a particular fly. Noël Deerr, the world's most prolific expert on sugar cane (he devoted two volumes to the subject published in 1949 and 1950), discovered a beetle that is indigenous to New Guinea and is parasitised by a fly. The beetle feeds on sugar cane and cannot survive without it. The fly cannot live without the beetle. For such an exclusive relationship to have evolved, Deerr reasoned, sugar cane must have been established in this region for thousands of years.

There are hundreds of varieties of sugar cane growing wild in New Guinea, but until relatively recently it was thought that there was just one: *Saccharum spontaneum*. The cane that actually conquered the world originally also evolved in New Guinea – the Creole cane (*Saccharum officinarum*). This cane would eventually be described by the famous botanist and founder of the modern taxonomic system of classification Carl Linnaeus in 1753. *Officinarum* means 'of the apothecaries' shops', since sugar was, for a great deal of its history, considered to be a medicine and was thus sold in the medieval equivalent of a pharmacy. The phrase 'an apothecary without sugar' came to describe a state of utter helplessness.

We now know that it was this Creole cane that the Polynesians subsequently carried with them on their migrations, chewing on the stalks to boost their energy. They took it to Indonesia, the Philippines and the northern tip of India around 8,000 BC. Later it travelled to Fiji, Tonga, Samoa, the Cook Islands, Marquesas, Easter Island and Hawaii. The Hawaiians offered it to the Sun God and carried it with them as they migrated northwards. From Hawaii sugar cane was taken to the rest of India, but when this occurred is unclear.

The earliest reference to sugar itself is in *Patimoksha*, the first record of the Buddhist life, which may have been handed down orally from the Buddha himself. *Patimoksha* said that, whenever the Buddha is not sick, he 'shall partake of delicacies to wit, ghee, butter, oil, honey, *gur*, fish, flesh, milk curds'. *Gur* is the Indian word for sugar. The Buddha (*c.* 563–483 BC) grew up and taught in a region where sugar cane was cultivated and so his life and teachings have become enmeshed in a sticky web of sugar.

What is extraordinary about sugar in India is that somehow sugar cane became synonymous with both major religions, Hinduism and Buddhism. Why this occurred is difficult to tell: it may be because the Buddha lived in a sugar-cane-growing region, or perhaps because sugar-cane cultivation evolved in tandem with the growth of religion in India. Despite what we now know about sugar, it is a force for good and a positive influence in these religions. Sugar must have seemed luxurious, a small indulgence in a hard life, and an even sweeter gift for both Hindu and Buddhist monks and nuns following vows of piety, poverty and chastity.

It may have been a linguistic misunderstanding that led to the link between Buddhism, Hinduism and sugar cane. There are numerous legends that are common to both faiths of a mythical kingly race of India, descended from Ikshvaku, founder of the solar dynasty, and himself the son of Manu, father of all humankind. The Sanskrit word for sugar cane is *ikshu*, which has no connection to Ikshvaku, but the words are so similar there seems to have been some confusion and, in later tales, the Ikshvaku race became intertwined with the origin of sugar cane.

In one tale, a priest tells the story of a rich and powerful king, Subandu. One morning he discovered that a sugar cane had begun to grow in his bedroom. The king was perplexed about the cane and consulted his Brahmins who told him that it was a good omen and a cause for rejoicing. The king left it alone and the cane flourished. It grew and grew until one night, while the king and his queen slept, the cane split apart to reveal a baby. Surucira, the queen, accepted the baby as her own, and the following day a ceremony was held to celebrate the arrival of the new prince. On the advice of the Brahmins, Subandu and Surucira called the prince Ikshvaku, which means sugar cane. Ikshvaku had many descendants, the hundredth of whom was the Buddha.

This legend appears in the *Mahavastu*, in which the life and teachings of the Buddha are related, and is just one of many that refer to the origin of the Buddha and the importance of sugar cane. One of the legends in the *Mahavastu* tells of the Buddha, at the end of his seventh week after enlightenment, resting under a tree by the roadside. Two merchants, Trapusa and Bhallika, were passing and recognised that he was a holy man. They gave him a piece of peeled sugar cane. This joint of cane was the first food he'd eaten since enlightenment and, as a result of their kindness, Trapusa and Bhallika became the first official Buddhists.

Another tale is of Ajatasastru, King of Magadha, who gave the Buddha a field of sugar cane. But one day a man was leaving this field with a bundle of cane and refused to give any to a child who asked him for some. In his next existence he was condemned to be a ghost in a luxuriant field of sugar cane that beat him to the earth every time he attempted to eat it.

One of the signs that a Buddha has descended are that ghee, honey, molasses and sugar never run out in the house where he lives. The Buddha used many metaphors derived from sugar: he said that his teachings were as full of sweetness. The Buddha's son, Rahula, had a foster mother who wanted to become a nun, but the Buddha would not allow women to enter any religious orders. She followed him from Kapilavastu to Vesaili and pleaded with him. Ananda, favourite disciple of Sakya Muni, supported her petition. The Buddha relented but said, 'Just as when the disease known as *manjitthika* falls on a field of ripened sugar cane, that field does not last long, even so, Ananda, in whatever discipline of *Dhamma* women are allowed to go forth from the home to the homeless life, that godly life will not last.' *Manjitthika* means the 'colour of madder' and is a red dye still used in India today. The most widespread sugar-cane disease is red rot; it is caused by the fungus *Colletotrichum falactum*. Red rot is prevalent in Vesaili; its devastating effect only becomes apparent as the cane is about to be harvested. This is very likely the first description of the disease.

In the Indian state of the United Provinces, the Hindu festival held in the middle of November that celebrates the god Vishnu awaking from four months of sleep has become a celebration of sugar. Traditionally the zemindar, who plants the sugar cane, worships Hindu gods in a field of the crop. Five girls collect sticks of sugar cane: five of these canes are placed on the eastern edge of the field, and five canes each are given to the priest, the blacksmith, the carpenter, the barber, the washerwoman and the landowner. The landowner arranges his five canes in a pyramid round a footprint on the floor to symbolise the god Vishnu. The women sing songs to raise him from his sleep and a priest then recites a prayer: 'Rise, O God, you lord of the universe, from your sleep. By awakening, the three worlds will also awake.' The priest then describes the miraculous origin of sugar cane and the subsequent birth of the Buddha. Feasting and festivities follow and the harvest can then begin.

The earliest references to how sugar cane was processed come from Buddhist works. A sugar mill is mentioned in AD 100; in AD 500 the various operations carried out in a sugar factory are used as a metaphor to teach Buddhist maxims. In another analogy, the Buddha describes what we now call 'bagasse': 'As a sugar-cane stalk is thrown to the ground to be dried for burning after all the juice has been extracted by pressing in the mill, so the body pressed in the mill of old age awaits the funeral pyre.' However, sugar cane was generally processed by being crushed in a mortar and pestle. Even today in some parts of rural India, the only technological innovation has been the creation of a giant mortar and pestle turned by an ox.

Even at this early stage in the development of sugar, there were several varieties of sugar cane. In the *Charaka-Samhita*, a medical textbook written around AD 78 by Charaka, the Kashmiri physician to a Buddhist king in Peshawar, there is a reference to a white cane called *paundraka*. The cane was named after the region it came from: Pundra, in northwest Bengal and northeast Behar. Originally the people, the Paundrakas, were of a military origin but had fallen from grace, becoming a lowly caste known as the Puds or Pods. They were the world's first professional sugar boilers. The region's name, Pundra, derives from the Sanskrit root meaning to pound or reduce to powder: now Pundra is described in Sanskrit dictionaries as 'the country of the sugar cane'. According to Deerr, 'There appears therefore just a possibility of the country having been named from the fact of its possessing sugar-cane plantations.' He goes on to say that there is a place in Bengal called Paundra-vardhana where the Paundras lived and the cane flourished leading him to suggest that Bengal was the first place in India where sugar cane was cultivated. Furthermore, in Bengal there is a city, now ruined, called Gaur, a very similar word to that for the most common kind of sugar sold in India – *gur*.

In spite of primitive extraction systems, India perfected sugar refining very early in its development and they had names for several different grades of sugar: *phanita*, a concentrated syrup; *matsyandika*, which was partially solid; *guda*, solid but amorphous, sold as *gur* or jaggery today; *khanda*, a low-grade white sugar; and *sarkara*, a higher grade of white crystals. The best-quality sugar was called *sitopala*.

So, it is to India that we most likely owe our word for sugar. The Sanskrit word *çarkara* was derived from the Pakrit *sakkara*, which originally meant sand or gravel, and later became *sukkur* in Arabic. The Sanskrit *khanda*, meaning to break, has become candy.

Ancient texts, such as the *Arthasastra*, which probably dates from AD 100–200, give explicit instructions on how to grow cane:

> Sugar-cane plants are propagated by cuttings, which are plastered at the cut end with a mixture of honey, clarified butter, the fat of hogs and cow dung ... lands in the neighbourhood of flooded areas are best for long pepper, grapes and sugar cane. Always, when sowing seeds a handful of seeds bathed in water with a piece of gold shall be sown first, when this mantra call be recited: 'Salutation to God, Pajapati Kasyapa. May the Goddess Sita flourish in my seeds and gods.'

Until the 1800s most physicians, like Charaka, believed that sugar was a medicine. He wrote, 'Wine made from sugar is agreeable to the mouth, is a slight intoxicant, fragrant, destructive of all diseases of the anal canal, aids digestion and when old promotes cheerfulness or relish and improves the complexion.' He also thought it promoted sperm production.

However, sugar wasn't only treated as a medicine. The Indian epic the *Mahabharata* mentions sugar including punch made from sugar, lemon juice, water, spices and rum (our word 'punch' comes from the Indian word for five; there are five ingredients in this traditional punch). The Indians also made wine from sugar and date palms, and to some concoctions *dhataki* flowers or fruit juice were added. The Chinese traveller Hiun Tsang observed in AD 620, 'They feed themselves generally on cakes of parched gram, which they mix with milk, cream, butter, solid sugar and mustard oil.' Later the great Arabic traveller Ibn Battuta visited India during his sojourn in the East (1325–54) and wrote that at a fête in Delhi he was given an unusual drink: 'They offer cups of gold, silver and glass, filled with sugar-water. They call it sherbert and drink it before eating.' The sherbert, he said, was flavoured with essence of rose petals.

In the *Arthasastra* the author provides instructions for the Superintendent of Elephants: 'The rations of an elephant shall be a strengthening drink of ten *palas* of sugar and one *adhaka* of liquor, or two *adhakas* of milk.' And under the section on 'Wonderful and delusive contrivances', the author has the following advice: 'The scum prepared from a mixture of the root of *kasruka* [a water creeper], *utpala* [sugar cane], and mixed with *bisa* [water lily], *durva* [grass], milk and clarified butter will enable a man to fast for a month.'

It was in India that Westerners received their first taste of sugar. The earliest Western written reference to sugar cane dates from 325 BC when Nearchus, an officer in Alexander the Great's army, which invaded the Punjab, described eating a milky rice pudding sweetened with sugar. This is still a popular dish in India today. It contains pudding rice coated in melted margarine with spices such as cinnamon, cardamom and cloves, as well as sultanas, desiccated coconut and possibly almonds, all simmered in milk with a fair amount of sugar. When I was a child my mother used to make this same dish for me but using rice vermicelli. Although it has a proper Indian name, *seviyan*, we used to call it string pudding.

Another mention of sugar cane was from the Chinese Buddhist pilgrim Fa Hien, who entered India east of the Indus in AD 399. He wrote, 'As you go forward from the mountains, the plants, trees and roots are all different from those in the land of Han, except the bamboo, the pomegranate and the sugar cane.' This indicates that sugar cane was also well established in at least parts of China. It had been taken there, possibly along with the process by which it was manufactured, by the Indian Buddhists in around AD 50. About this time Buddhism split into two – the Mahayana, the Greater Vehicle, and Hinayana, the Smaller Vehicle – and the Mahayana sent missionaries to China.

One Chinese story described by Li schi tsching, author of an encyclopaedia, *Pen ts'ao kung mu*, tells of an Indian Buddhist, Zou, who had settled at See tschuan and lived in a hermitage. He owned a white donkey and when he wanted provisions he sent the donkey to market with some money and a note asking for salt, rice, firewood or vegetables. The local people would attach his goods to the donkey and send it back up the mountain to him. One day his donkey got into a field of sugar cane owned by a Mr Huang and created considerable damage. As compensation Zou showed Mr Huang a superior method (to the Chinese one) of how to make sugar from his cane. At this time the Chinese made *shimi*, which literally means 'stone honey'. They squeezed the juice from the sugar cane and boiled it, but it was still thick and dark with molasses. Zou's process created a finer, whiter sugar.

The Emperor T'ai Tsung is alleged to have sent a deputation to Behar in AD 640 to learn the art of sugar boiling and, as a result, it was announced that Chinese sugar was superior to Indian sugar. Even so, many years later, when Marco Polo was travelling through China during the fourteenth century, he wrote of Fukien (now Fujian):

> They have an enormous quantity of sugar. From this city the Great Kaan gets all the sugar that is used at Court, enough to represent a considerable sum in value. You must know that in these parts before the Great Kaan subjected it to his overlord-ship [Gengis Khan's Mongolian invasion], the people did not know how to prepare and refine sugar, as is done in Babylon. They did not let it congeal and solidify in moulds, but merely boiled and skimmed it, so that it hardened to a kind of paste and was black in colour. But after the country had been conquered by the Great Kaan, there came into the regions men of Babylon [Egypt] who had been at the court of the Great Kaan, and who taught them to refine it with ashes of certain trees.

Because sugar cane had been brought to China by the Buddhists, it is no surprise that sugar continued to be associated with Buddhism in China. Fasting monks and nuns would drink sugared water and there was a sugar ceremony to worship the Buddha on his birthday, the eighth day of the fourth month. Water fragrant with herbs was boiled with sugar; called *yu Fo shui*, it means 'bathing the Buddha water'. Buddhist icons were carried in a procession by monks and nuns and were then soaked in the sugared water. This water was distributed to the onlookers for blessing themselves. The ceremony derives from the belief that when the Buddha was born the heavens opened and he was showered with sweet, scented water. In the eighth century a procession took place in the city of Loyang where a thousand Buddhist statues were carried through the city, watched by over a million people. When one considers that sugar was eaten widely in the monasteries – at this time there were over a quarter of a million monks and nuns and several hundred thousand novices in the country – and used heavily in this ceremony, China must have required an enormous amount of sugar.

At this time it was still only the religious orders and the elite who had access to sugar, the latter eating honeyed bamboo shoots, honeyed ginger, pickled crab and a cake made from sugar, rice powder and milk. The rest of the population could not afford sugar. But, by the time of the Song Dynasty in the early thirteenth century, sugar had become a luxurious necessity, even for Chinese peasants. The novelist Wu Zimu, in *The Past Seems a Dream*, described a night market in Hangzho that had no less than seven shops specialising in sugar. They sold honey cakes, flower-shaped candy, sweet rice porridge, spun sugar, sugar pastes, musk-flavoured sugar, preserved sugar – in total 37 varieties of cakes, candies and syrups made from sugar. They made a kind of marzipan from ground pine nuts and sugar and used it to press into moulds to create sugar flowers. Edible sculptures of spun sugar in the shape of lions, birds, flowers and fruit were also common. Teahouses sprang up, and with the tea they served wine, pickled vegetables, salted melon seeds, and fruit preserved in sugar. Elaborate porcelain jars filled with preserved fruit became fashionable gifts among the Song urban elite.

It was at this time that the Chinese started making 'rock candy' by boiling sugar-cane juice until it was supersaturated and thick. It was dried in the sun before being packed in jars and called *Tang shuang* – sugar frost. In Li schi tsching's earliest work, the *Tao hung king*, written during the Liang dynasty between AD 502 and 560, he described six different kinds of sugar and sugar products: *Shimi* and *Tang shuang*, as well as *Tschi tang*, a thick, brown juice; *Panmi*, concentrated syrup; *Sch tang*, sugar clarified using milk; and *Scha tang*, sand sugar (it's possible that the latter was derived from the Indian word *sakkara*). He also listed four different kinds of cane: the *Tschu tsche*, a soft green cane with very sweet juice; *Ti tsche*, the western cane, which may mean that it came from India; *Tek sia* or *La tsche*, the wax cane, referring to its thick rind; and the *Hung tsche* or *Tsie tsche*, a brown or purple cane that, he said, 'can only be enjoyed when eaten. Sugar is not made from it. If the juice expressed by a mill is drunk, it is pleasant enough, but not so delicious as when the cane is eaten whole.'

Sugar was also carried by the Indian Buddhist missionaries to Japan in AD 755. By the eighteenth century sugar was being manufactured by Japanese clans, the most famous of which was the Takamatsu who made Sanuki sugar, known to the west as Sambon White.

But the most important introduction by the faithful was from India to Persia, now Iran, 600 years after the birth of Christ.

Though there is little other written evidence, documents show that a tax levy was imposed on sugar cane cultivated in Mesopotamia during AD 636–44, so sugar cane must have arrived in Persia prior to this time. It's again mentioned in an account of the capture of Dasteragad, a palace near Baghdad used as a residence of the Persian king Chosroes II, by the Roman Heracliys in AD 627. One Persian tale refers to his predecessor, Chosroes I, who lived during the sixth century. The Sultan was passing a beautiful garden when he saw a girl and asked her for a drink of water. She brought him a cup of sugar-cane juice cooled with snow. It was so delicious he asked her how she made it. She said that sugar cane grew so well in her garden she could squeeze the juice from it with her bare hands. He asked her for another cup and, when she was gone, he thought to himself that he must remove the girl and her family and make the garden his own. To his surprise the girl returned without his drink and burst into tears. When he asked her what was wrong, she said that his intentions towards her had changed. She said she knew this because she had been unable to wring a single drop of juice from the canes. Chosroes realised the injustice of his thoughts and gave up his idea of owning the garden. He told the girl to try again and this time she quickly came back with a full cup of juice. This story is repeated in the *Arabian Nights*, but Hollywoodised – the Sultan was about to impose a tax on anyone who could grow that much sugar; in the end, he marries the girl.

The Persian sugar industry survived until 1300. The last Sultan, Mostasim, was executed by the Mongol Hulagu, Ghenghis Khan's grandson, in 1258 when his army captured Baghdad. The collapse of the empire followed and the sugar-cane industry subsequently disintegrated.

One hundred years after sugar cane had reached Persia, it spread to Egypt and flourished alongside the Nile. By the tenth century it was an important crop, growing along the Persian Gulf, the southern shore of the Caspian Sea, in Mesopotamia (now Iraq), the Damascus oasis, the valley of Jordan and Egypt.

Despite this expansion, it was an event that happened when sugar cane was first introduced to Persia that led to the world domination of sugar cane – the Arabian army's quest to convert the world to Islam. In their conquest of Persia they came across sugar cane and it seems likely that, from then on, sugar cane was transported throughout the western world by the Arabic Islamic faithful.

What made sugar so special? A number of crops came from India and spread to the Mediterranean via the Middle East at around the same time as sugar cane: cotton, bananas, mango and taro, and yet we don't have an economy fuelled by mangoes. In addition, sugar cane is difficult to grow: it needs a lot of fertiliser and huge amounts of water (28 irrigations per crop in Egypt; even cane grown in areas that were not as hot still required large amounts of water); it can take eighteen months to mature and, as a result, doesn't fit into any crop rotation schemes so cannot be grown with other food crops. It was plagued by hordes of voracious mice and caterpillars and could be devastated by dogs, elephants and hippopotamuses. Above all, it was growing at the limits of its tolerance in the Mediterranean where the summers were too dry and the winters were too cold.

Professor Jock Galloway, who now teaches Geography at Toronto University, Canada, and has become an expert on sugar cane, believes that initially 'the barrier to the diffusion westwards from India was not just environmental; it was political, cultural and institutional'. Galloway's introduction to sugar cane was a somewhat unusual one. 'One snowy March morning many years ago, Professor Theo Hill of the Department of Geography, McGill University, stopped me . . . to ask me if I'd like to spend the coming summer at McGill's Bellairs Research Institute in Barbados. How could I refuse?' Once on Barbados, Galloway discovered that his responsibilities – monitoring some climatological instruments – took only a few minutes a day, and so he found his way to libraries and archives and into the world of the sugar-cane industry. Referring to the plant's spread to the Mediterranean, he adds, 'The creation of the conditions in which these agricultural improvements could be introduced was of crucial importance to the timing of the westward diffusion of crops out of India and it is the singular contribution of the Arabs to have brought these conditions about in the years following the birth of Islam.'

There were three main factors that led to the spread of sugar cane by the Arabs and all were inextricably linked: the Arabs brought with them prosperity, peace and education. As Galloway says, sugar-cane production needed:

> landowners and labourers to learn new agricultural techniques; the construction of irrigation schemes required a large investment of capital, a legal framework to govern the distribution of the water and a market for produce. The market could come from rising prosperity and an increasing population which in turn depended on peace and political stability ... Missionary fervour and military conquests that within a century had brought new religion and Arab rule to a broad swath of territory from Sind to Spain gave way to a kind of pax Islamic, in which the unity of language, religion and sometimes even of administration made for the relatively easy movement of people and goods and knowledge. Traders and scholars, pilgrims and governors, all sorts of people travelled to and fro between ports and caliphal capitals observing and learning as they went.

The Arabs fostered a respect for education. They cultivated beautiful gardens and undertook agricultural experiments. Their scientists analysed the chemical process needed to solidify and refine sugar; they came up with a way of making sugar last longer and travel better. The Arab conquest also removed restrictions against change, such as exempting sugar cane from taxation. One poet at the time wrote: 'Where the Arabs set foot on Spanish soil life and water sprang up, the sycamores, pomegranates, bananas and sugar cane intwined in the glistening labyrinths, and even the very stones blossomed in gay colours.' An Italian traveller, the author of *Viaggio a Gerusalemme*, wrote of Cyprus in the fifteen century: 'The abundance of the sugar cane and its magnificence are beyond description ... In this area it is impossible to believe that anyone can starve. It is charming to see how the best qualities and the inferior grades [of sugar] are made, and how the people, nearly four hundred, are employed.'

Another scholar, Andrew Watson, who specialises in agriculture under Islam, writes, 'The Islamic contribution was less in the invention of new devices than in the application on a much wider scale of devices which in pre-Islamic times had been used only over limited areas and to a limited extent.' In order for sugar cane to prosper, engineers had to adapt and distribute the water supply. The water wheel was known but not widely used in pre-Islamic Middle East: the Arabs improved the design and enabled its geographic range to be extended. Likewise the Persian (Iranian) invention of tapping the water table by means of an underground canal had limited distribution until the Islamic conquest.

By this stage, the Arabs understood the basic principles of sugar-cane growing. The plant is asexually propagated, which means that sections of the cane can be planted and will grow into a new cane. This first crop is harvested by cutting the cane back to a stump; each cane will then regrow a second and a third time. These secondary and tertiary crops are known as ratoons. 'It is necessary [for the sugar cane] to have low, sunny land which has water near at hand,' wrote Ibn el Awam, a twelfth-century Arab chronicler of agricultural practices. He was describing cultivation in Spain.

It is planted from its own roots, and also from its own cane, having made the soil previously very mellow with . . . a good layer of good, light, decomposed manure, and dividing into squares of ten codos long and five codos wide [one codo equals 45 centimetres] . . . they are irrigated every fourth day and when grown to the height of a hand are cultivated well, and manured with a heavy layer of sheep dung and continued irrigating every eight days until the beginning of October, from which month irrigation is stopped.

The canes were at risk from a number of pests. In Egypt special walls of clay and chopped straw were built with overhanging parapets to baffle the mice. To get rid of caterpillars, the cane was watered with tar. Frederick III of Sicily had a different approach: he commanded his subjects to each collect up a certain number of caterpillars every day, 'on which means one can have more reliance than on the prayers of priests, trailing over the fields'.

There was, however, little technological innovation as far as processing the cane to extract the juice was concerned. Mills designed for flour, to crush grapes or extract oil from olives were used. Normally they consisted of a rotating grindstone over an immobile stone, or a round grindstone sitting upright and turned with a horizontal crankshaft by a person or an animal. Any final extraction was carried out with a press – a series of wooden boards in which the cane was inserted and screws were tightened.

A writer, Carusio, described a factory in sixteenth-century Sicily. He said that, when well-refined sugar was required, the syrup was boiled three times, and after each boiling it was poured into the cone-shaped clay jars. Excess syrup drained through a hole in the bottom and the white residue left behind was the refined sugar. Sugar produced in this way was known as loaf or cone sugar. Milk, eggs, alum, wood ashes and lime were all added to the syrup as it was boiling to help remove the impurities and clarify it. The surplus juice was often fed to horses.

A contemporary description of a Sicilian factory was rather more colourful than Carusio's:

> Near to Palermo . . . there is a beautiful, diversified and delightful plain, adorned with vineyards and fertile fields and especially abounding in sugar cane . . . There are also on this plain buildings called *trapeiit*, where the sugar is solidified. If one enters one of these it is like going into the forges of Vulcan, as there are to be seen great and continuous fires, by which the juice is solidified. The men who work are blackened with smoke, are dirty, sweaty, and scorched. They are more like demons than men.

Egyptian sugar was mainly exported, usually to Venice. Different grades of sugar were named after the country or town of origin: Alexandria, which had a reputation for quality, was from Egypt; there was also Bambillonia, Cairene, Caffetino, Damaschino and Muscavado. The latter was made in large three-kilo loaves, rounded at the top, and was popular among dealers as it could be broken into saleable quantities. Essence of violet or rose water was added to give a pale-pink 'rose sugar'. Candy, molasses, sugar crystals and syrups in glass jars from Alexandria were also exported.

Sugar became widely used in feasts and festivals and by rich households. The Vizier Qafur household consumed 454 kilos of sweetened foods a day in AD 970. A Persian traveller, Nasir-i-Chosran, recorded the Ramadan feast of the Fatimite Khalif, Motansir, in 1040 where fifty thousand men of sugar were paraded for the people, each one weighing nearly two kilos. The most opulent feast of the time, though, was held when the Nile was allowed to flood into specially built canals.

Sugar was distributed among the guests; depending on their rank they received from half a kilo to twelve kilos of sugar each. Over a hundred cooks used just over twenty tons of sugar to make such delicacies as hens boiled in syrup and figurines of elephants, lions, giraffes, deer, plus two huge castles crafted out of sugar. At the end of the feast, beggars were invited to carry them away. Although the poor did not usually manage to eat so richly, they could probably afford poorly refined sugar and molasses.

The Arabs from the Arabian Peninsula, a huge area now encompassing Saudi Arabia, Yemen, Oman, United Arab Emirates, Qatar, Bahrain and Kuwait, conquered vast tracts of the Mediterranean and North Africa. They swept through Persia, Egypt in AD 640, Cyprus in AD 644 and Sicily in AD 655. In AD 682 they took Morocco and the rest of North Africa; Spain fell in AD 711, Crete in AD 823 and Malta in AD 870. As they conquered these countries, they took sugar cane with them, introducing it to Europeans on a wide scale; their plantations, factories and refineries flourished from AD 700 and lasted for almost a thousand years.

However, it was during this flourishing period that a crucial event took place, which was to change the world politically, culturally and economically.

There is a reed, from which flows a very sweet juice, called cannamelli zachariae; this honey they eat with bread and melt it with water, and think it more wholesome than the honey of bees. Some say that it is the honey of Jonathan, the son of Saul, found in the earth and disobediently tasted. With the juice of this reed our men, at the siege of Albarria, Marra and Archa [Acre] often stayed their hunger.

So wrote the soldier Jacobus de Vitriaco, a Crusader, who came across sugar cane in Syria around the turn of the twelfth century. The Crusaders had set out to recover the Holy Land, Palestine, from the Muslims, but along the way they encountered sugar and sugar cane. The first effect of this discovery was that the Crusaders became responsible for the first proper introduction to England of sugar, which up until then had been a matter of hearsay and second-rate rumour; this led to a taste for sugar that has grown exponentially from that time to this.

The English king, Henry III, asked the Mayor of Winchester to procure him three pounds [1.4 kilos] of sugar, including violet and rose sugar, in 1226. During 1288 the Royal Household used 2,722 kilos of sugar at an exorbitant cost, equivalent today to nearly £60,000. It would be another five hundred years before sugar was cheap enough for the average English person to afford it.

This first taste of sugar in western Europe led to a growth in demand that stimulated expansion in cultivation in Palestine, and the development of industries in Rhodes, Malta, Crete and Cyprus as the Crusaders conquered these lands. In the early fifteenth century, the Genoese, encouraged by the Portuguese monarchy, tried to establish sugar in the Algarve; by the 1450s it was growing as far north as Coimbra. It was planted in Tuscany in the 1550s, and in the 1560s and 1570s Catherine de Medici tried to cultivate sugar cane in her gardens at Hyéres in Provence.

The second consequence of the Crusades, as a result of this expanding demand, was to forge a link between the sugar industry and European colonialism. The Arabs had brought it, but it was cultivated under Christian rule in Cyprus, Crete, Rhodes and Greek Morea. The Arabs used slaves; for example, in Morocco, they coerced Greeks, Bulgarians, Turkish prisoners of war and Tartars from the Black Sea. But what the Christians established was an embryonic form of the plantation system that was to be modified in the Atlantic islands of Madeira and São Thôme, and would be put into practice with ferocious vigour in the colonies.

For a plantation system to work it needed four things: wealthy overseas financiers, a well-advanced manufacturing industry, an expanding market in Europe and, of course, almost free labour. The Crusader orders, such as the Knights of St John, acquired extensive properties and financial muscle. They oversaw the creation of a sugar-cane industry that was based on exporting sugar to Europe, was dependent on sophisticated and well-established refining industries (mainly based in Venice), required a high capital investment, that the Knights, naturally, could supply, and used serfs, prisoners of war and slaves. In effect, Crete and Cyprus became colonies operating under Venetian rule between 1204 and 1489. A not atypical family, the Cornaros, of Venetian origin based at Episkopi between the mid- and late fifteenth century, employed four hundred labourers, most of whom were slaves.

Helmut Blume, author of *The Geography of Sugarcane*, writes that the Cornaros' plantation was large, but others, though smaller, were of the same type. He adds:

[The plantations] had a specific social structure and settlement pattern where the production was capital-intensive and export-oriented. It is precisely this type of agricultural-industrial enterprise which became the classical colonial plantation as sugar cane spread from the Mediterranean to the Atlantic Islands and then to the European colonies in tropical America. Thus the origins of the plantation system were laid by Europeans in the Levant and the Mediterranean islands late in medieval times.

In Salobreña, on the southern coast of Spain, a few kilometres from Motril, a Moorish castle appears to grow out of a large rock, encircled by narrow, winding streets and tiny, white houses whose window boxes are choked with cacti and bougainvillaea. It's marooned in a green sea of sugar cane that stretches in one smooth wave down to the beach. This is practically all that is left of Europe's once-glorious fields of cane. In the deserts of Morocco are the remains of sugar factories, crumbling into the sand – the ruins of what was a flourishing industry. But by the end of the sixteenth century, sugar production in the Mediterranean had all but collapsed. The traditional reason given is that sugar started to be produced in colonies overseas and the competition undercut Mediterranean sugar. In practice, the situation was more complex.

The decline had begun a century earlier due to the lethal combination of war and plague. In 1324 Egypt had 63 refineries in Fustat, now the capital, Cairo; a century later only nineteen were operating. In the second half of the fourteenth century war enveloped the Crusader states in Syria and Palestine; they were then invaded by the Mongols and, in the fifteenth century, the Ottoman Turks. Under Turkish rule, Europe's economy collapsed and its population was devastated by a series of plagues during the fourteenth and fifteenth centuries. By the fifteenth century the population had been reduced by a third. Poor maintenance of irrigation and fluctuations in the Nile led to crop failures, the death of cattle and famine. The poverty-stricken rural peasants fled to the city seeking medicine and money. Plantation owners sought to recoup money and replace their dead workers with slaves as free labour and the Genoese aristocracy pushed to claim colonies in the Canaries and on Madeira.

But something else contributed to the end of the sugar-cane industry in the Mediterranean – climate change.

Sugar had been grown in the Mediterranean during a warm period, but between 1550 and 1700 there was what was known as the Little Ice Age. The climate was cooler and less rain fell. Sugar cane struggled to survive and yields dropped. It is likely that, in time, the sugar plantations around the coast would have collapsed anyway, but the final blow came when sugar from a new Portuguese colony, Brazil, began to arrive at a much lower price than it could profitably be produced in the Mediterranean. Although Brazilian sugar was handicapped by high transportation costs, producing it was cheaper: it needed no irrigation; there was an abundance of land and the colony relied extensively on slavery.

Once the Mediterranean market disintegrated, sugar cane spread to South America, the Canaries and the Caribbean; finally sugar cane crossed to the east coast of America and to the other side of the world – to Australia. But, before sugar could reach South America, a stepping stone on the way towards sugar cane's global domination was established in the Atlantic: the island of Madeira.

2. CHRISTOPHER COLUMBUS AND THE DRAGON TREES

Like amber presented on a crystal plate,
Is there anything more beautiful than crystallised sugar?

Su Shi, Chinese poet

To men who had spent months at sea, sighting any land at all would have been welcome, but they must have been overjoyed to see this particular island rise jaggedly from the ocean. As the seas calmed off the island's southern shore, until the sun's reflection lay like a slick of molten butter on the water's surface, they would have been able to sail close enough to observe Madeira's outstanding beauty. A volcanic outcrop with coal-black lava beaches, sheer cliffs and deep fissures as if the land had been chiselled open, it was covered in dense forest and riven with streams. The year was 1421 and the discoverers of this balmy island that lay off the western coast of Morocco were men who owed their allegiance to Prince Henry of Portugal, famous for his maritime navigational skills and his devotion to the sea and its study.

Madeira was a vital stepping stone from the Mediterranean to South America and the Caribbean, from small fields of sugar cane to the full-blown plantation system that resembled a factory more than a farm, and from hired labour to the acceleration of slavery and the African slave trade.

In 1421 Henry divided the island into two; he gave the northern part to one of its co-discoverers, Tristão Vaz Texeira, who founded the city of Machico, and the other half to Gonçalves Zarco, distantly related to the Portuguese royal family, who helped to create Funchal, the capital. In addition, he presented an island off the coast of Madeira, Porto Santo, to a young nobleman, Bartolomeo Perestrello, then only twenty. Perestrello had served both his father and Henry well, in spite of his youth – or so Henry wrote when he drew up the deeds to the island. (Perestrello's daughter would marry Christopher Columbus, and the Perestrello family name was invaluable to Columbus.) Henry sent Malvosie grapes from Crete and sugar cane from Sicily to aid the colonists. But Madeira's first and best export was timber. The sheer volume of wood changed the very style of architecture in Portuguese houses; even the design of their ships improved. The island was so densely wooded that the colonists had to start a fire to clear enough land for crops and it is said that it raged for seven years.

The first sugar cane was planted by Texeira. In 1432 Prince Henry entered into a contract with his squire, Diogo de Teive, and the two erected a water wheel and built a sugar mill; the crown was to receive a third of the proceeds. Eventually the new colony grew enough cane to influence the course of the European sugar trade: the first consignment to arrive in England docked in Bristol in 1456. By the end of the century the island was churning out the maximum amount of sugar that it could: 1,790 tons a year, making it the largest producer of sugar in the western world. Sugar was the bedrock of the island's economy: Funchal's municipal seal was formed from two canes complete with roots and leaves, and in between and on either side of them were conical loaves of sugar.

Prince Henry was also responsible for starting the trade in human lives. In 1441, twenty years after he'd founded Madeira, the prince returned from an exploration of the coast of West Africa with Muslim captives of mixed Arab and African blood. These men said they were a proud race and unfit for bondage, but in the interior of Africa lived the accursed children of Noah's son Ham (Ham's sons were said to have fathered the southern peoples of Africa, but Noah had cursed Ham and one of his sons, Canaan, saying that their descendants would be 'servants of servants'). In return for their freedom, they offered to capture them. So, in 1444 they received their freedom and 235 Negroes were shipped out of Lagos, the first of millions who were to lose their lives to sugar.

Later the Portuguese crown permitted cane growers to send a ship every two years to Guinea to collect slaves. From 1567 the island was authorised to import 150 slaves a year for 5 years. Slaves were mainly used to clear land and create ditches for the new crop. Though these figures were low in comparison to later sugar colonies, Madeira thus continued the link between sugar and slavery established by Christian Europe, and marked the beginning of the African slave trade.

In 1478 Columbus arrived on Madeira. He was about 27 years old, and no one has a clear idea of what he looked like (all surviving portraits of him were painted after his death) or his true origins. It is thought he was born in Genoa, the peasant son of a wool comber, but he had already begun to learn his future trade as a sailor by the time he was fourteen. Contemporaries describe him as a tall man with flaming red hair, blue eyes and freckles; he was charismatic and an eloquent speaker with a good sense of humour and a terrible temper. He was to let nothing stand in the way of his dream of conquering the ocean.

Columbus had been sent to Madeira by a businessman, Paolo di Negro, to buy 2,400 *arrobas* – about 36,000 kilos – of sugar. However, it is alleged that, although di Negro had been given the full amount of money by his business partner, he did not pass it on to Columbus. Columbus had loaded his boat with sugar in Funchal before being told by the port authorities that the city did not extend credit. While negotiations took place, Columbus stayed in the city centre with a wealthy businessman, João Esmeraldo, for six days. Later he would be a welcome, if infrequent, visitor at Esmeraldo's house. The house itself has been destroyed but in 1989 the site was excavated by a team of archaeologists, who found a well and many shards of pottery, from plates to sugar moulds to vessels for holding olive oil, as well as coins, clay pipes and a pig's molar.

The 1478 dispute was only resolved by unloading part of the cargo, but Columbus appeared in court in Lisbon a year later for not bringing the full amount back. Fortunately, he was not prosecuted, particularly as he'd told the court he could only spare them one day. It is not known what was so important that legal proceedings could not detain him, but it may have been his wedding to Filipa Moniz Perestrello, Bartolomeo's daughter. He had seen her in church and sent her a note asking her to meet him secretly afterwards. She was not a beauty and it is unclear how deeply Columbus was in love with her.

The two met a few times before Isabel, Filipa's formidable mother, heard of their illicit arrangements and demanded to see Columbus. The meeting was satisfactory to both of them. Perestrello had died a few years before and the family was relatively poor so Isabel was pleased and relieved that Columbus, though he earned little himself, did not ask for a dowry. Columbus, however, above wealth, longed for nobility. As the son of a peasant it was difficult for him to attract the attention of the kind of people who had the financial means to send him on business ventures such as di Negro's, or to fund longer, more explorational voyages. The name Perestrello and the connections his future family possessed would be the making of Columbus. So obsessed with nobility was he that, after Filipa's death, when he was 37, he fathered a son with a twenty-year-old girl, but refused to marry her because she was a peasant. Throughout his life he invented seals and signatures that suggested he came from a noble lineage in decline and which have plagued historians attempting to divine his true origins.

When Isabel Perestrello agreed that Columbus could marry her daughter, Columbus also gained something else: Isabel consented that, as soon as they were wed, Columbus could have the keys to Perestrello's library. Perestrello had been one of the few men who had explored parts of the Atlantic. In addition to his own navigational charts he had amassed maps and information in a private collection that was only rivalled by the one Prince Henry had created at the School of Navigation he had founded and which Perestrello had attended.

Porto Santo occupies a tiny 45 square kilometres, 37km from Madeira. It has twin volcanic peaks connected by a sandy plain. The beach, stretching in front of Vila Baleira, the main town, is breathtaking – an unbroken line of golden sand, book-ended by sheer cream-and-russet cliffs. And that's it. Apart from a few houses, Porto Santo has nothing: few resources, not even much water, no agriculture and, apart from grass, little in the way of vegetation. For this Porto Santo has Bartolomeo Perestrello and his colonists to thank.

When Perestrello set sail from Portugal to claim ownership of his island, he took with him a pregnant rabbit that gave birth on the ship. Perestrello saw it as a good omen. Unfortunately, on their arrival, the rabbits bred madly, eating all the local plants as well as those the colonists tried to grow. Later Columbus was to say that he would never eat rabbit again after he left Porto Santo. Originally the island was forested, although not as densely as Madeira. It also had groves of a strange tree that wept blood. Only a handful now remain in the central plaza of Vila Baleira in front of the site where Christopher Columbus's house once stood – they may even have been there when Columbus was alive. Some of these trees had such wide girths that the trunks could be hewn into canoe-type boats for up to seven men. The gum was brilliant red and resinous; the colonists sold it for lacquer. This and the multiple crowns of thick, fleshy leaves lent it its name – the dragon tree. They'd grown for centuries before the colonists all but wiped them out.

Perestrello lasted two miserable years before heading back to Lisbon. Henry persuaded him to return and agreed to pay him an annual sum during his lifetime. After his death, Isabel, who had always hated the desolation and lack of culture on Porto Santo, and who now had little money, sold the island. When her son, also called Bartolomeo, came of age, he sued for possession of the deeds to the island. He won his case and his mother was forced to pay back her buyer. This was the situation in which Columbus found himself: living on an unprosperous satellite colony of Madeira in a small house with his new wife, her mother, and his brother-in-law Bartolomeo and his wife. As well as the tensions between Bartolomeo and his mother, there were other strains: Columbus is likely to have spent more time with Isabel than Filipa would have liked, for she was more knowledgeable; she had, after all, catalogued her husband's library. Both Filipa and Bartolomeo were resentful of Columbus, whose frequent travels selling merchandise meant that the care of his new baby son, Diogo, and his wife, who had become increasingly ill after the birth, fell to his brother-in-law and his wife.

However, in other ways Porto Santo was a formative place for Columbus. As well as studying his father-in-law's maps, he paced the island, particularly the beach, attempting to work out what was west of where he was. Why, he wondered, did the winds and currents from the Atlantic transport driftwood that had clearly been weathered for years, and also brought the seeds, birds and insects that had created Madeira's rich fauna and flora, which had not come from the east, from Africa? It is allegedly here, on the long and perfect beach at Porto Santo, that Columbus formulated his plan. He reasoned that if he were to sail west across the Atlantic he would travel right round the world, arriving in the Orient, in China and India.

Although Filipa was ill, Columbus still undertook long voyages; after one trip he returned to find that her mother had taken her to see specialists in Lisbon. Columbus travelled there, but she died a few days after his arrival in Portugal, of tuberculosis. Columbus left the Perestrello family and Porto Santo and, with his brother (also confusingly called Bartolomeo), he petitioned the crown of Portugal, King John II, to give him the money to sail westwards to China. King John was cordial each time he was approached by the

Columbus brothers, but his council of mathematicians decided on Columbus's first visit that the voyage was foolhardy and thus could not possibly be profitable.

The second time Columbus spoke to the king, he believed that John was raising money and a crew who would undertake the same journey but without Columbus. He was a paranoid man anyway, so it is unclear whether there was any truth in this. Nevertheless, Columbus, his brother and Diogo, who was four or five at the time, sailed secretly to Spain in 1485.

It took eight years to persuade King Ferdinand and Queen Isabella to sanction his scheme. The love and faith of his brother and Beatriz de Arana, the peasant girl with whom he had another son, Hernando, kept Columbus sane. He could not have lasted so long without some other more material form of encouragement and it is said that Queen Isabella, although never sanctioning his project (indeed, she rejected it a number of times), was more supportive than her husband, who became increasingly preoccupied with war against Portugal.

There were many reasons why it was so difficult to convince Spain of the virtues of his enterprise. Partly there was Columbus himself, a bright autodidact but certainly no scholar, a foreigner with an ambiguous past and a peasant mistress. There was the war, and, although many, by this stage, believed the world was round, the idea of circumnavigating the globe to reach India and China was novel. Columbus calculated the distance and the time it would take before he, too, could follow where Genghis Khan had once led his hordes. There was no way he could have guessed that another continent lay across what was referred to as 'The Darkness' – the other side of the Atlantic. Indeed, the year he eventually sailed *Hortus sanitas* was published, lavishly illustrated with sea monsters, such as the Marine Turk and Marine Virgin. Even in 1598 in *Das Fischbush* (The Book of Fish), there was a drawing of a sea demon, 'captured in the Adriatic at the beginning of the fifteenth century'.

Father Hernando de Talavera was one of the key men who had advised the King and Queen of Spain against Columbus's scheme but, on hearing that Columbus was finally about to abandon his hopes of Spanish help and petition France instead, he rethought. Although he believed the project would fail, if Columbus succeeded, he did not wish him to bring riches and glory to another power. And so

Columbus was finally given royal blessing. The money to fund his trip was raised by the crown and Genoese businessmen and bankers.

In 1492 Columbus set sail, leaving behind his mistress and their son, to whom he would not return, as well as his first son, Diogo.

The first substantial landmass Columbus encountered was Cuba, which he rather optimistically thought was China. On 12 December 1492 he discovered Hispaniola (later to become St Domingue and then Haiti). Hispaniola would be the next great step in the journey of sugar cane. Here it would be argued over by the religious; it would dramatically alter the landscape and civilisation, lead to the deaths of thousands of indigenous peoples and herald the rise of the slave trade. On 18 February 1493, aboard his boat, Columbus wrote the first official description of the island for his patrons:

> Española is a wonderful island, with mountains, groves, plains and the country generally beautiful and rich for planting and sowing, for rearing sheep and cattle of all kinds and ready for towns and cities . . . the rivers are plentiful and large and of excellent water; the greater part of them contain gold . . . The people of this island and of all the others which I have discovered or heard of, both men and women, go naked as they were born . . . They have no iron or steel nor any weapons . . . they might be made good Christians and be disposed to the service of Your Majesties and the whole Spanish nation and help us to obtain the things which we require.

Columbus left both sugar cane and colonists on Hispaniola before sailing back to Spain in triumph. He brought with him six captured 'Indians' (so called because of his belief that he had reached Asia). So that they would not be lonely, he kidnapped seven 'wives' and three children, though it is likely that there was only one complete family on board. Isabella was horrified. She had the Indians baptised, gave them lodgings and issued a royal decree that they should not be harmed. To her credit, in the next few bloody decades she acted as honestly, reasonably and ethically as it was possible for her to do in the face of what would become a wholesale slaughter of these people and a burgeoning trade in human lives sanctioned by her husband, the king.

Columbus was given a hero's welcome. There is one story of a banquet held in his honour. A guest remarked that there were other men who could navigate, sail and had an understanding of mathematics and cosmology. He implied that Columbus was undeserving of so much approbation. Columbus asked for an egg to be brought. When it arrived he bet that no one could make the egg stand on its pointed end. The noblemen tried in vain, passing it round the table. When it returned to Columbus, he smashed the end; the egg would now stand upright on the flattened shell. The point, he said, was not that there were no other bright or skilled men in Spain, but that no one but he had thought of doing what he did or undertaking what he had undertaken.

When Columbus returned to Hispaniola in 1494, the colony he had founded had been abandoned and the Indians had fled. The fortress the early settlers had built had been burned to the ground and only the charred remains were left. It took some time to find the Indians and win back their trust. When he finally did so, he discovered that, once he had sailed for Spain, the Indians said, the colonists led a debauched life. They stole gold from the indigenous people and raped the women. They pushed further into the interior of the island searching for gold and entered the territory of another tribe. The king of this tribe and the adjoining one joined forces and gathered together a band of warriors who marched on the Spaniards and killed them all.

Columbus was forced to send an abridged account back to Spain. On 30 January 1494 he wrote to Ferdinand and Isabella to say that, although the canes had rooted, the accompanying men had sickened and died:

> We are sure by our trials that in this land the vine and wheat will grow well; it is to be expected from the speed of the growth of the vines, wheat and sugar-canes that have been planted, that the products of this place will not be behind those of Andalusia and Sicily . . . May it please your Highnesses to instruct Don Juan de Fonesca [the Bishop of Burgos] that he send only the best sugar-canes and he must see that the canes sent are of good quality.

He also asked Don Juan to bring him fifty casks of molasses from Madeira 'as it is the best sustenance in the world and the most healthful . . . and also ten cases of sugar, which is very necessary'.

Ferdinand and Isabella endorsed his requests.

From now on, Madeira's sugar domination began to wane, partly because of its geography. Madeira's volcanic peaks are exceptionally steep. The island still retains many of the original terraces that were hewn into the hillsides – they are barely more than a few metres wide and shored up with dry-stone walls. The new colony owned by the Portuguese in Brazil – sugar cane had arrived in Brazil in 1500 and the first shipment of sugar arrived in Lisbon in 1526 – soon undercut Madeira's prices. For a while Madeira imported the worst sugar Brazil produced, refined it and shipped it back to Europe, but eventually this, too, became uneconomical.

Today there are but a few small terraces of sugar cane in Canhas, west of Funchal; the stems, tied in fat bundles to protect them from the wind, are purple with a glaucous bloom like that of a plum. A little further along the coast, at Calheta, are the ruins of a sugar mill, the chimney crumbling and the giant iron machinery lying half-submerged in the sand. A few metres away is one of only two remaining working sugar mills on the island. Once a year Madeira's tiny crop of cane is crushed between giant iron rollers; the juice is distilled to form a thin syrup and molasses. The syrup is used to make *aguardente*, a kind of rum. The molasses, called sugar honey by the locals, is added to almonds, dried fruit and mixed spices to make *bolo de mel*, a flat, sticky cake that tastes like gingerbread. The terraces are instead filled with vegetables, bananas and bird of paradise, their spiky blooms grown as a cut-flower crop. Only a few trees remain, gathered along the cloud line.

Madeira has, instead, taken advantage of Prince Henry's other legacy: his vines produced the first wine for export in 1445. By all accounts it was an average table wine. Fortunately, it has not remained that way. In one shipment destined for the East India Company not all the wine was sold and the story goes that, on its return, it had improved substantially. At first the Madeirans thought the transformation was wrought by the rocking motion of the ship and sent their wine on long sea voyages; now they know the secret is warmth, Madeira is made with the addition of brandy and is gently heated for up to six months before being aged in a wooden cask. The resulting four types of Madeira are a cross between sherry and port, with notes of hazelnut and caramel. *Sercial* is dry and served chilled, as an aperitif; *verdelho* is medium dry; *boal*, medium sweet, frequently accompanies strong cheese such as Stilton, and *malmsey*, the sweetest, works best with coffee and desserts. Winston Churchill is said to have drunk a bottle in Funchal from 1792 and commented, 'To think that this wine was made when Marie Antoinette was alive.'

After Columbus had founded Hispaniola he proceeded north, excitedly renaming the islands he encountered: Trinidad, San Salvador, Dominica and St Iago (or St James, now Jamaica). He took cane from Hispaniola to Tenerife, Palma and Grand Canary towards the end of the century, but did not succeed in establishing a viable industry due to hordes of voracious caterpillars and competition from Portugal's new colonies.

The rest of his stay in Hispaniola was ignominious to say the least: in 1499 he was deposed as viceroy and sent back in chains by his successor Bobadilla. He was restored to favour by Ferdinand and Isabella and allowed to return to Hispaniola, but not as viceroy. Bobadilla was recalled to Spain, though he died in a hurricane on his way home; Columbus had predicted it, but his reputation was so tarnished no one had listened. Although now believed to be one the world's finest navigators, Columbus ended his life in obscurity, living in Seville on the gold he had stolen from Hispaniola.

Initially Hispaniola must have seemed like Eden: there was no need for the planters who'd followed Christopher Columbus to spend money on irrigation systems, terraces or fertiliser – when one field of cane was harvested, they simply cleared more land. It took sugar cane a while to become fully established in the New World, mainly because of the lure of gold as an alternative source of revenue, but the first New World sugar did come from Hispaniola: six loaves were presented to Charles I of Spain (later Charles V, Holy Roman Emperor) in 1516. The first West Indian sugar to reach England also came from Hispaniola, a cargo that arrived in Bristol, bought in exchange for North African slaves financed by six wealthy Englishmen.

By the mid-sixteenth century sugar cane counted for nearly 70 per cent of exports from Hispaniola, but, not long after this, production plummeted. The gold was nearly gone, and the Indian population had dramatically declined from eight million to only sixty thousand. By 1520 their genocide was complete.

The Spanish king entrusted the governing of Hispaniola to three Hieronimite friars, Luiz de Figueroa, Alonza de Santo Dominigo and

Bernandino Mazanedo, in the hope they would help save the Indians – and the economy. A legal adviser, Alonzo de Zuazo, also accompanied the friars. Meanwhile Spain moved the bulk of her sugar-cane interest to Cuba, Jamaica and Mexico. The three friars tried to build up a new economic base, exporting maize, manioc, cotton and hides to Europe. The Cassia tree, also known as the golden shower or the pudding pipe tree, was introduced and the friars exported it as a type of medicine. Unfortunately, there was a cassia glut on the market and the situation started to seem hopeless.

Shortly afterwards, the three were followed by another Bartolomeo, a priest who had a partner in the sugar business in Cuba. Unhappy with the treatment of the Indians in 1517 Bartolomeo put forward a scheme whereby each Spaniard living in Hispaniola could import twelve African slaves. It is thought that he meant Spanish-born Africans but De Zuazo wrote directly to Ferdinand asking for permission to import Negroes from Africa, and it was his proposal, not Bartolomeo's, that was accepted. The initial number agreed was 4,000 slaves, thereby establishing the sugar industry. De Zuazo also anticipated the triangular slave trade by bartering articles from the Cape Verde islands for slaves. Later Bartolomeo became known as Las Casas, the Apostle of the Indies. But he was equally horrified at the trade and use of African slaves and felt responsible. Before his death he wrote, 'It is as unjust to enslave Negroes as Indians and for the same reasons.'

De Zuazo set up a scheme whereby entrepreneurs could be loaned up to 500 gold pesos to build sugar mills. He himself constructed one of the largest. In 1518 he wrote of 'glorious fields of cane as thick as a man's arm and twice as high'. It was an expensive business and became a plaything of the elite: Genoese aristocrats supplied money for three of the most productive mills. Charles I of Spain provided capital, too. One of the finest sugar mills, built at the mouth of the Rio Niazo so that sugar could be shipped directly from the factory, was called Nueva Isabella. Nueva Isabella was owned by Diogo Colon, High Admiral of Spain and Viceroy of the Indies, who held court with his royal bride, Maria de Toledo, grandniece of King Ferdinand of Spain, in the palace he had constructed in the jungle. But all that is left of Nueva Isabella today are stones and lizards.

In 1580 an epidemic killed most of the slaves on Hispaniola. In addition, the King of Spain by then ruled Portugal so he had access to

Brazilian goods and Spanish markets as well. The prevailing wind and ocean currents meant fleets loaded with gold and silver from Mexico and Peru could leave the Caribbean via the Straits of Florida; commercial and naval traffic no longer stopped at Hispaniola, preferring to dock at Havana instead.

In an effort to prevent contraband trade in hides, the friars forcibly relocated the indigenous population to the eastern third of island, and sugar plantations in this vicinity ceased production. Runaway slaves desecrated the remaining plantations.

According to sugar historian Jock Galloway, Hispaniola represented just a transitional stage in the evolution of the industry from the small-scale farming of the Mediterranean to the large plantations in the Americas. The gradual increase in slaves on the island culminated in the industry's dependence on slavery, which lasted until the nineteenth century.

Although production on Madeira was low, during the peak years it produced more sugar than Hispaniola, leading Galloway to conclude that in Hispaniola sugar was had become the business of a few rich entrepreneurs. There is also the frequent allegation that Spain was the first country to begin the slave trade, importing slaves from Africa on a regular trade route. Sidney Mintz, professor of anthropology at Johns Hopkins University, Maryland, and author of *Sweetness and Power* says, 'It was Spain that pioneered sugar cane, sugar making, African slave labour and the plantation form in the Americas.'

As for Portugal, when her colony Madeira fell into decline, and Spanish Hispaniola struggled to found its sugar-cane industry, Portugal created what was to become one of the most powerful sugar-producing empires of the Americas: the interim colony of São Tomé followed by the conquest of Brazil.

São Tomé is a volcanic island off the coast of West Africa in the Gulf of Guinea. It was discovered in 1471 by the Portuguese explorers Pedro Escobar and Jaõa Gomes. Except for a century of Dutch rule between 1641 and 1740, Portugal controlled the island until its independence in 1975 and its inhabitants are almost solely descendants of the original slaves shipped there by the sugar-cane industry.

According to Galloway, it was here that Portugal finally abandoned Mediterranean practices and wholeheartedly adopted a new approach: large plantations worked by African slaves. By 1550 it was producing 2,203 tons of sugar a year, and by 1570 it had 120 sugar-cane mills.

Though the sugar could be sold at a higher price once it had been refined, it was usually sold before it was refined at a lower price than either Brazilian or Madeiran sugar. The discrepancy may be due to the difficulty in drying the sugar in such a tropical and humid climate, but Galloway believes it is more likely that the plantation owners were responding to the preference of refiners, who made a greater profit by importing cheap sugar and refining it.

Almost from the beginning São Tomé operated on a different scale from Madeira or any of Spain's new acquisitions in the Canaries. By 1550 there were 6,000 slaves on the islands; a wealthy man now owned 300 slaves whereas prior to this a sign of affluence was to have fourteen. Portugal sent their undesirables to this island – including 2,000 Jewish children, in the hope they'd become Christians. As early as 1517 these different conditions created a slave uprising. Further damaging revolts occurred in 1580, 1595 and 1617; Portugal lost control of the island briefly in 1580, and, when the Dutch initially attacked in 1598, they laid waste to the plantations.

Madeira had needed terraces carved into its steep hillsides and constant irrigation. It lay far away from the European markets, and was riddled with problems: once the trees had been cleared soil fertility declined dramatically and the cane was ravaged by disease, rats and insects. São Tomé on the other hand benefited from its tropical climate and its ability to export low-grade sugar for the foreign refineries. While it lasted, it brought Portugal considerable profit.

Brazil was sighted in 1500 by Pizon, Columbus's pilot. The following year, on Good Friday, Captain Cabrall landed – he took possession of Brazil for the Portuguese crown on Easter Monday. Any nobleman who wanted a share in the country was given a stretch of coastland 240 kilometres long and absolute power over his land. The first sugar from Brazil was brought by a navigator, Martin Affonso da Sousa from Madeira, in 1532; barely a year later he established a factory in São Vincente. The Jesuits built two factories a short time later in Rio de Janeiro and, for the rest of the century, Brazil continued to enable her masters to prosper with more sugar and of a better quality than the West had yet seen. Thanks also to Madeira and São Tomé, Portugal had become the world's largest sugar producer and seemed indomitable. The Portuguese had learned how lucrative sugar could be and had realised – unlike the less successful Spanish, whose colonies were dogged by dreams of gold and silver – that there was no better investment. As Noël Deerr eloquently put it, they knew that 'no Eldorado glittered over a western horizon to tempt settlers to abandon the plantations and mine for gold'.

The Portuguese and the Spanish were thus the first slave traffickers, but they were arguably more humane than other Europeans. They allowed slaves to marry and discouraged the break-up of marriages by the sale of one party. Generally slaves were allowed to make some money with which they could buy their freedom, and they were converted to Christianity. The Bishop of Luanda in Angola, for instance, would sit on an ivory chair on the quayside, baptising bound slaves as they rowed beneath him on the way to the slave ships. Though they were, of course, being converted against their will, it meant that the Spaniards and the Portuguese viewed their slaves as human beings with souls.

Once the New World had been discovered, any unknown lands were arbitrarily divided between Portugal and Spain by Pope Alexander VI. The dividing line, created at the Treaty of Tordesillas in 1494, was the meridian 1,786km west of the Cape Verde Islands; it almost coincides with the fiftieth meridian west of Greenwich. At the beginning of the sixteenth century Spain owned Hispaniola, Cuba, Puerto Rico, Jamaica and Trinidad as well as the whole of South America west of the Tordesillas Line, while Portugal had Brazil, Madeira, São Tomé plus lands to the east of the Line. Between them, the two nations controlled the whole of the Americas. The result was that indigenous populations were virtually wiped out. It is estimated that forty to fifty million people lived in what is now the Spanish-speaking Americas and 90 per cent of them died, partly as a result of brutality but mainly due to their lack of resistance to diseases the Spanish and Portuguese brought with them. The death toll did not end there. Between 1500 and 1650, Spain and Portugal exported half a million African slaves.

But the pattern of world history and the path of sugar cane were about to change. The French, Dutch and English were on the verge of fighting for their own colonies and the result would be a radical shift in the balance of power leading to a huge increase in the amount of land devoted to sugar cane and an exponential rise in the number of men, women and children enslaved to sugar.

The French were the first to challenge Spain. One of the most sensational attacks was made by Jean d'Anjo, a merchant from Dieppe, who hijacked a Spanish galleon in 1523 and seized 80,000 pesos' worth of gold bars, crown jewels, royal dresses stiff with pearls and emeralds as well as a substantial proportion of sugar. The English and the Dutch joined in the skirmish soon afterwards. The English Crown, in effect, sanctioned piracy as a legitimate means of seizing wealth and intimidating other powers. Sir Francis Drake, the first Englishman to circumnavigate the globe, led a campaign against Spain between 1577 and 1580 in his ship, the *Golden Hind*. He pillaged the coasts of North and South America and attacked Spanish fleets; on his return to England in 1580 laden with treasure, he was duly knighted by Elizabeth I. The Dutch pirate Piet Pieterzoon Hein captured the whole of the Spanish treasure fleet with booty valued at fifteen million guilders in the Mantanzas Bay of Cuba in 1628.

The French, Dutch and English changed from consumers of highly refined sugar produced by Spain and Portugal to colony owners and sugar manufacturers themselves. Lured by the false promise of easy money, these three nations initially took over unoccupied islands: France snatched St Kitts, Martinique and Guadeloupe, and England seized Barbados, an island in the Caribbean overlooked by Columbus. At first they tried to grow other cash crops, such as cotton and tobacco, but, when these did not prove as lucrative as they had hoped, they turned to sugar.

The Dutch had won and lost Brazil and now owned Guiana. Spain's power gradually diminished – in the Treaty of Madrid signed in 1670 she formally abandoned her claims to Jamaica and the Cayman Islands to Britain. The three remaining powers, the French, Dutch and English, fought over their colonies and many islands changed hands several times over the next hundred years, with England eventually retaining Barbados and owning British Guiana, Jamaica, Grenada, St Vincent, Dominica, St Kitts, the Virgin Islands, Antigua, the Bahamas, Bermuda and Mauritius. France controlled Nevis, Tobago, Martinique, Guadeloupe, Reunion and St Dominique. Sugar made more for Britain than the combined profit of all her other colonial products; it was equal to half the value of France's.

By 1700 most of the Caribbean and the neighbouring South American mainland had become a sugar-producing area owned by North Europeans.

3. A TRUE AND EXACT HISTORY OF BARBADOS

> The condition of slavery has existed from the time when man first organised himself into communal societies, and has continued until quite recent times to form part of the social economy of those nations that pretend to civilisation.
>
> Noël Deerr, *A History of Sugar,* vol II, 1950

Governor James Kendall wrote in 1690 of Barbados, 'Itt is the beautyfull'st spot of ground I ever saw.' Barbados is indeed beautiful with its brilliant-white sandy beaches tinged with streaks of coral-pink fringed by causarina and palm trees. The eastern coast is a sweeping curve dashed by the Atlantic with limestone outcrops worn into tortuous shapes. The island is only 33.8km long and 22.5km wide but is densely populated with 260,000 people, most of whom live along the south and west coast. Hotels and apartment blocks line the thin spits of sand, cheek by jowl with barbecue fish stands and stalls hawking bananas and sweet potatoes. In spite of the encroaching golf courses, the inner section of the island is still dominated by fields of sugar cane, an undulating grassy lawn almost six metres high. There is almost no remnant of the original forests that covered the island. The Portuguese, when they saw it shrouded in the bearded fig festooned with whispering roots, gave the island its name: the bearded one.

Our introduction to Barbados in the seventeenth century comes from a somewhat reprobate character, Richard Ligon. Author of *A true and exact history of Barbadoes*, written in 1657, he described himself on the frontispiece as a 'Gent.', but proceeded to give a fuller account of himself, beginning the book:

> Having been censur'd by some (whose Judgments I cannot controll, and therefore am glad to allow) for my weakness and Indiscretion, that having never made proof of the Sea's operation, and the several faces that watery Elements puts on, and the changes and chances that happen there, from Smooth to Rough,

from Rough to Raging Seas, and High going Billows (which are killing to some Constitutions,) I should in the last Scene of my life, undertake to run so long a Risco as from England to the Barbadoes; And truly I should without their help conclude my self guilty of that Censure, had I not the refuge of an old Proverb to fly to, which is, [Need makes the old Wife trot] for having lost (by a Barborous Riot) all that I had gotten by the painful travels and cares of my youth.

This convoluted introduction serves as a justification for his situation at the time: Ligon lost his possessions in a riot that occurred during the post-Civil War turmoil in England and became a refugee. He struck up an acquaintance with Thomas Modyford, a Royalist exile, and sailed with him to Barbados on 16 June 1647. He arrived to discover the newly established colony sick with the plague, corpses piled by the roadside and the survivors suffering from a severe food shortage. Ligon helped Modyford build a profitable plantation and later went into partnership with an earlier settler, William Hilliard.

Ligon was wildly enthusiastic about Barbados, its food and, in particular, its women. He was captivated by Negro women – the first one he saw was the mistress of a plantation owner who entertained him and his companions on his arrival.

Dinner being ended . . . we made room for better Company; for now the Padre, and his black Mistress were to take their turns; A Negro of the greatest beauty and majesty together: that ever I saw in one woman. Her stature large, and excellently shap'd, well flavour'd, full ey'd, and admirably grac'd . . . her eyes were the richest Jewels, for they were the largest, and most oriental that I have ever seen.

Shortly after his arrival, he encountered two young black girls on a road. In semi-ecstasy he followed them and tried to offer them money to make them his mistresses. He described them in precise detail, particularly their breasts, which 'stand strutting out so hard and firm, as no leaping, jumping, or stirring will cause them to shape any more, than the brawn of their arms', before finally attempting to excuse his behaviour to his readers. Ligon proved to be as racist as most men of his era, finding it uncomfortable to see white servants working in the fields alongside black slaves and, after only three years, he returned to England to face arrest and imprisonment for debt. He does, however, provide one of the best eyewitness accounts of the early days of the fledgling colony.

Barbados was the beginning of English involvement in the Caribbean. Captain John Powell landed there in 1625 on his way back to England from Brazil. He set up a cross in St Jamestown, now called Hole, and carved on a nearby tree: 'James K of E and of this island'. Powell did not seem surprised that such a beautiful island was deserted – the indigenous population of Amerindians is likely to have been wiped out by the Spanish by the time they deserted the island in 1536, well before the English arrived. Barely any trace of their presence remains today: archaeologists have unearthed a few necklaces made of fine animal teeth, fishing hooks carved from whelk shells on lines woven of pineapple fibre and shards of pottery. As well as hunting and fishing, these first people grew cassava, yam, plantain and bananas.

Powell immediately saw the potential of the island and, on his return, spoke to the owner of his ship, Sir William Courteen, a Dutch merchant who had settled in England, about the possibilities. A couple of years later the two set sail for Barbados with eighty settlers and seven or eight Africans stolen from another ship. These men were the first slaves to land on an English settlement. On Powell's original journey back to England, he captured a Portuguese ship with a cargo of sugar that Courteen sold for £9,600 and donated to the new settlers. Powell is also said to have brought over thirty to forty Amerindians from Guiana to teach the colonists how to plant cane. By 1730 every Amerindian reintroduced to the island had died, mostly of maltreatment, malnourishment or disease.

Initially the colonists planned to grow tobacco, cotton, indigo, ginger, cassava, plantain, beans and corn; the cane was only used to make a refreshing drink and was not fully established until much later. This sugar cane is likely to have come from Madeira around the 1640s: a letter dated 3 January 1678 from the East India Company at Madras to Captain Robert Knox in Ceylon reads: 'One piece of sugar-cane carried by a Madeira ship to Barbados was the first cause of that great plantation and manufacture, and that within the memory of some living.' What the English did not realise was that sugar cane was already present on the island; it had probably been introduced by the Amerindians.

Richard Ligon described the island's first attempt at growing sugar:

> At the time we landed on this island, which was in the beginning of September 1647, we were informed that the great work of sugar making was but newly practised here. Some of the most industrious men have gotten plants from Fernambrock [Pernambuco, Brazil], made tryall of them, and finding them grow they planted more and more, until they had such a considerable number to set up a very small ingenio [sugar mill], and to make tryall what sugar could be made on that soyl. But the secret of the work not being very well understood, the sugars made were very inconsiderable, and little worth for two or three years ... At the time of our arrival we found many sugar-works set up and at work, but as yet the sugars they made were but bare muscovados and few of them merchantible Commodities, so moist and full of mollosses, and so ill cured that they were barely worth the bringing home to England.

Yet again, it was the religious who helped to establish this fledgling sugar-cane industry. To escape the Inquisition, many Sephardic Jews of Portugal had travelled to Amsterdam during the sixteenth century. Some converted and became 'New Christians'; others went on to Brazil. In both Pernambuco and Amsterdam they became involved in the sugar trade.

Jock Galloway writes, 'Given the close bonds of religion, business and family between the Sephardic communities of Pernambuco and Amsterdam, the Jews in Pernambuco became the allies of the invading Dutch and, as a consequence, on the defeat of the Dutch in Pernambuco [by the Portuguese], they had to emigrate once again.' Preferring to remain in the same industry, many sailed to the Caribbean where they advised French and English colonies. In Martinique, which had been colonised by the French, they were not made welcome by the governor, M. Du Parquet. In contrast, the governor of Guadeloupe realised that the Jews not only had knowledge, they had capital. In 1665 a M. Houel gave them land in perpetuity and supplied them with slaves, oxen and carts. In exchange they undertook the cultivation of 26.3 hectares of land, two-thirds the amount Guadeloupe's newly established mill could cope with. They were allowed to keep three-fifths of the profit; the rest plus the molasses was given to the governor. When Du Parquet realised his mistake, he invited the Jews back and offered them superior terms. One of those who returned to Martinique to take up Du Parquet's offer, Benjamin D'Acosta, not only had an influential role in organising the sugar industry, he also introduced coca to the island.

Ligon noted that the Barbadians had also taken advice and wrote:

They were grown greater Proficients in the boyling and curing and had learnt the knowledge of making [sugar] white . . . but not so excellent [as] those of Brazil, nor is there any likelihood that they can ever make such, the land there being better and lying on a Continent, must needs have a Constanter and Steadier weather, and the Aire much drier and purer than it can be in so small an Island as Barbadoes.

An anonymous account in 1679 also credits the Dutch for the success of sugar in Barbados:

The Dutch beginning to lose their footing in Brazil many came from those parts, who taught the English the Art of making sugar, and having at that time free trade with all people in amity with England . . . their sugar yielded a good price, and they were plentifully supplied with all necessities of life and planting at very cheap rates, and had long credit given them by the Dutch, which together with their being governed by themselves was the beginning and main cause of their prosperity, and grew and increased very much year by year in the production of their manufacture and in this condition they continued until about the year 1650.

It wasn't until 1655, however, when England invaded Jamaica, that sugar from Barbados began to reach the mother country – just over 7,000 tons in that first year. The first fifty refineries were opened in England. It marked the beginning of England as a serious sugar-producing empire.

In order to control the remaining Amerindians from other islands, Lord Francis Willoughby, the Governor of Barbados, appointed Indian Warner, the son of Sir Thomas Warner, Governor of St Kitts, and an Amerindian woman, to the post of Deputy Governor of Dominica in

1662. He was treated as a European while he was brought up in his father's household, at least until his father's death in 1648, and by Lord Willoughby. However, in 1666 he was captured by Du Lion, Governor of Martinique, and kept in irons as if he were a slave. Lord Willoughby's brother, Lord William Willoughby, eventually managed to secure his release and reinstated him.

The Amerindians had managed to retain control of some areas of Dominica and St Vincent from which they launched raids against the colonists. In 1674, Sir William Stapleton, who had become Governor-General of the Leeward Islands three years before, decided Indian Warner was implicated. The raiding parties were, he believed, Indian's private army and not warring Amerindians from French colonies. A punitive expedition was dispatched under the command of Philip Warner, Deputy Governor of Antigua and Indian's half-brother. Not realising the duplicity, Indian and his Amerindians helped Philip against the French Amerindians, and were then invited to a rum party. Once they were drunk, everyone was slaughtered, even Indian's children. One member of the expedition, William Hamlyn, reported Philip, who was arrested and committed to the Tower in England. But, at the trial, Hamlyn's evidence was discredited and he was not allowed to speak. Philip Warner was acquitted. Although Charles II removed him from his position and forbade him from ever holding office under the crown, he was regarded as a hero by the planters, and elected Speaker of the House of Assembly in Dominica.

Once the Amerindians had been wiped out, the plantations were run by indentured white servants – either prisoners or servants who had been kidnapped or seduced into accepting service without understanding the nature of their service. There was even a word for it: to be Barbadosed. Laurance Saunders, Mayor of Bristol, made a practice of persuading minor offenders to apply for transportation, sharing with the shipper the profits of their sale to the planters. He wasn't the only one: when the infamous Judge Jeffreys (known for his harsh sentences) arrived in Bristol in 1685 he found that many of the aldermen and justices of the city were occupied with this trade. An eyewitness account described how the judge turned to the mayor:

> accoutred with his scarlet and furs, and gave him all the ill names that scolding and eloquence could supply . . . he made him quit the bench and go down to the criminall's post at the

bar; there he pleaded for himself, as a common rogue or thief might have done; and when the mayor hesitated a little or slackened his pace, he bawled at him and stamping called for his guards.

As a result of the mayor and his colleagues' efforts, Sir Josiah Child was to write in 1718:

Virginia and Barbados were first peopled by a sarte [sort] of lose vagrant People; vicious and destitute . . . either unfit for Labour, or had so misbehaved themselves by Whoreing and Thieving or other debauchery that none of them would set to work . . . gathered up about the Streets of London and other places, cloathed and transported to be employed upon the Plantation.

Irish Catholics were frequently Barbadosed. One Irishman was given 21 lashes for remarking during dinner 'that if there was so much English Blood in the tray as there was Meat, he would eat it'. For the Irish and other indentured servants, there was often cruel treatment, as Ligon testifies.

As for the useage of the servants it is much as the master is, merciful or cruel. Those that are merciful treat their servants well both in their meate, drink and lodging, and give them such work as it is not unfit for Christians to do. But if the Masters be cruel the Servants have very miserable and wearisome lives . . . Their cabins are made of sticks, and Plantaine leaves, under some little shade that may keep the rain off . . . Their suppers being a few potatoes for meate, and a little Mobbie [beer brewed from sweet potatoes] for drink . . . They are rung out by a bell to work at six o'clock in the morning with a severe overseer to command them till the bell ring again, which is at eleven o'clock, then they return and are set to dinner with a mess of Loblollie [cold corn porridge], or Potatoe. At one o'clock they are rung out again to the field, there to work till six, then home to a supper of the same, and if it chance to rain they have no shift and must lie so all night . . . I have seen an overseer beat a servant about the head with a cane until the blood came, for a fault which is not worth speaking of.

Just before Ligon arrived in 1647 an island-wide servant rebellion was quelled and eighteen conspirators were executed.

By 1683 there were 2,381 indentured servants in Barbados; 10,000 had been shipped out of Bristol alone for the West Indies and Virginia. In Barbados they were called redlegs because of the ease with which these white men, particularly Scots in their kilts, burned in the fierce sun. Those who managed to survive their period of indenture scraped a living from the poor soil on the eastern coast of the island, referred to as the Scottish district, because its limestone escarpments and miniature mountains reminded them of home.

One man who was transported to the West Indies went on to become a plantation owner; his descendants were one of the wealthiest families to make money from sugar and the slave trade. In 1685 Azariah Pinney participated in the Duke of Monmouth's rebellions. Monmouth was Charles II's illegitimate son, and he unsuccessfully challenged his uncle, James II, for the throne. Pinney was on Monmouth's side. He was tried by Judge Jeffreys, who would normally have sentenced him to death, but, owing to his youth and his relatively wealthy background (his father ran a lace factory), he was only sent to the Caribbean. He settled in Nevis where he gradually accumulated enough wealth to buy a share in a plantation.

Other white workers were not as lucky. When most of them finished their contract, they were forced to leave Barbados, as the island was too small to allow the majority to buy their own land. Instead, Barbados shifted swiftly towards the full-blown plantation system: large areas of land almost solely devoted to growing sugar cane, with a few white landowners and a large number of slaves. Sugar was almost entirely exported while daily food as well as other goods had to be imported. The landowners became some of the richest people in England at the expense of black Africans: slavery increased from 6,000 slaves in 1643 to almost 40,000 within forty years. The irony is that England in the seventeenth century was evolving towards 'free' labour, where men and women worked for a salary, while enslaving men and women in the colonies.

Helmut Blume writes:

The intensive use of cheap labour, particularly black slaves . . . resulted in a specific settlement pattern and a special social structure, causing a rigid social stratification and racial tension.

It is for this reason that many scholars consider the colonial sugar plantation, as it developed in the American tropics, as an economical and political institution rather than merely a specific type of farm unit.

Sidney Mintz expresses it even more strongly, describing the plantation system as 'an absolutely unprecedented social, economic, and political institution, and by no means simply an innovation in the organisation of agriculture'. It was, he concludes, a unique combination of factory and farm.

The plantation system rested entirely on the slave trade. But how was the West able to procure so many slaves? Slavery already existed in Africa so it was, initially, a relatively straightforward matter for Europeans to join in the trade where human lives were simply another commodity. However, slavery in Africa and slavery in the plantations were quite different.

In Africa there were two types of slaves. Some slaves were loyal servants, including administrators and soldiers, and could become very powerful. Others were victims of slave raids; they were captured and used as labourers. Some well-to-do families were able to raise ransom money to buy back their relatives, as described by Baba of Karo, a member of the Hausa tribe from Nigeria in her eponymous book. Her aunt Rabi was kidnapped by African slave traders for the King of Abuja but was eventually rescued by her family. Captured slaves were frequently able to own their own land and could marry. Others, usually women, became the concubines of wealthy men. If they had children, they and the child acquired status. In many small communities the children of slaves were freed and the slaves and their families merged into the local community within a generation or two.

The Europeans' demand for slaves was insatiable, however, and greatly exacerbated and increased the existing system of slave raiding. Middlemen were employed who raised the money for gangs of men to bring slaves to the coast from the heart of Africa. This trade created new tensions as the European traders sought to deal directly with the groups in the interior to cut out the coastal middlemen and ultimately helped lead to the colonisation of Africa by the West.

One unnamed slave whose story was reported is not untypical. He or she said, 'They sold us for money, and I myself was sold six times

over, sometimes for money, sometimes for a gun, sometimes for cloth
... It was about half a year from the time I was taken before I saw
white people.' This slave would finally have met white men at the
notorious trading posts the Portuguese established along the west
coast of Africa in 1444. They built the first of the infamous slave
houses in 1536 on Goree Island, three kilometres off the coast from
the capital, Dakar. The Dutch founded the last in 1776. For 250 years
this island, now designated a UNESCO World Heritage Site, was a
slave depot for the millions of Africans who were deported to the
colonies. The main slave house is now a museum. Up to twenty men
were housed in cells 2.6 square metres in size, chained to the walls by
their neck and arms. Women, children and girls were housed sepa-
rately. Girls were valuable and were assessed by the state of their
breasts and whether they were still virgins. Some became the property
of slave traders who released them as soon as they became pregnant; if
the girl survived, she was referred to as *Signare*, a deformation of the
Portuguese *Segnora*, and her child became of a higher status than the
black Africans. A similar situation took place in the French West
Indies with the Creoles. There was a separate cell where men who
weighed under sixty kilos were held. Labelled 'Temporarily Unfit',
they were fattened, like geese, on the starchy beans the Senegalese call
niebe. Before they were sold, the slaves were paraded in front of the
traders – who actually slept in rooms above their cells. The slaves were
then forced through a thin corridor and out on to the docks. Some
tried to escape: if they were not shot by the guards, they were eaten by
sharks. The coastline was notorious for its shark population since sick
slaves were fed to them.

What followed was arguably even worse. As one would expect,
few slaves were able to learn to write. One notable exception is
Olaudah Equiano. Equiano was born in 1745 in an Ibo village he
called Essaka that lay in the interior of modern eastern Nigeria. He
was captured as a child and sold into slavery in Barbados. A
Lieutenant Michael Henry Pascal bought him and gave him the
name Gustavus Vass. At the time most slave owners called their
captives after Western heroes or famous English sites. While
travelling with Pascal, Equiano learned how to read and write and
later was able to attend school in England. Tragically, Pascal tired of
his pet and, when Equiano was twelve, he sold him to a Captain
Doran. Equiano, whose book, *The Interesting Narrative of the Life of*

Olaudah Equiano, or Gustavus Vass, the African, was published in his 40s, describes the experience:

> Captain Doran asked me if I knew him; I answered that I did not; 'Then,' said he, 'you are now my slave.' I told him my master could not sell me to him nor to any one else. 'Why,' said he, 'did not your master buy you?' I confessed he did. 'But I have served him,' said I, 'many years, and he has taken all my wages and prize-money, for I only got one sixpence during the war; besides this I have been baptised; and by the laws of the land no man has a right to sell me.' And I added, that I had heard a lawyer and others at different times tell my master so. They both then said that those people who told me so were not my friends; but I replied – it was very extraordinary that other people did not know the law as well as they. Upon this Captain Doran said I talked too much English; and if I did not behave myself well, and be quiet he had a method on board to make me.

Later Doran sold him to a Quaker, Robert King, and Equiano worked as a clerk for him in Montserrat. He was eventually able to buy his freedom. He frequently travelled on ships as a sailor, took part in the Phipps Expedition to the Arctic in 1772–3, toured the Mediterranean as a gentleman's valet and spent six months assisting an English doctor among the Miskito Indians of Central America. In the 1780s he became the leading African abolitionist in London, aiding other members of the Anti-Slavery Society, such as Granville Sharp, and acting as principal spokesman for fellow ex-slaves in Britain. His book was reprinted nine times in England alone and was one of the principal weapons that was used to repeal the Slave Act. An extraordinary man and an eloquent writer, it is thanks to Equiano that we have some idea of the horrors of slavery.

Once slaves had been bought, they were shipped from Africa to the colonies. The journey was called the Middle Passage and represents one of the most inhumane acts of cruelty perpetuated in human history. It took six to eight weeks, but could stretch to thirteen in bad weather. Equiano wrote of the experience:

> The stench of the hold while we were on the coast was so intolerably loathsome that it was dangerous to remain there for any time,

and some of us had been permitted to stay on the deck for fresh air; but now that the whole ship's cargo were confined together, it became absolutely pestilential. The closeness of the place, and the heat of the climate, added to the number in the ship, which was so crowded that each had scarcely room to turn himself, almost suffocated us. This produced copious perspirations, so that the air soon became unfit for respiration, from a variety of loathsome smells, and brought on a sickness among the slaves, of which many died, thus falling victims to the improvident avarice, as I may call it, of their purchasers. This wretched situation was again aggravated by the galling of the chains, now become insupportable; and the filth of the necessary tubs, into which the children often fell, and were rendered almost suffocated. The shrieks of the women, and the groans of the dying, rendered the whole a scene of horror almost inconceivable.

At least a third of the Africans died en route; those that survived often suffered from severe malnutrition, as well as profound psychological trauma. To increase profit, ships were overcrowded. A system was devised whereby, in contemporary parlance, they could be packed as tight as 'books on a shelf'. In the hold, below deck, slaves lay next to one another, chained in rows. Men had a space 180cm by 37cm and women had 160cm by 37cm, less room than if they were in a coffin. White servants were not given much more latitude, although they were often allowed a bed. A petition to Parliament in 1659 describes how 72 servants had been locked below deck during the whole voyage of five and a half weeks, 'amongst horses, that their souls, through heat and steam under the tropic, fainted in them'.

Air circulation was poor and they were kept in unbearable heat with inadequate supplies of food and water. Lack of toilet facilities meant they had to lie in their own waste and vomit: seasickness was highly prevalent. Faecal contamination and poor food led to outbreaks of typhoid, measles, yellow fever and smallpox. In addition to this degradation, many Africans imagined that they were being taken to Europe to be eaten, turned into oil or gunpowder, or that their blood would be used to dye Spain's flags red.

The slaves were often treated with great brutality and the women raped. John Newton, a slave trader who later repented of his previous life, reported the following incident in his log for Wednesday 31 January 1753:

> In the afternoon while we were off the deck, William Cooney seduced a woman slave down into the room and lay with her brutelike in view of the whole quarter deck, for which I put him in irons. I hope this has been the first affair of the kind on board and I am determined to keep them quiet if possible. If anything happens to the woman I shall impute it to him, for she was big with child. Her number is 83.

Few captains had such qualms; as Newton later testified, they were willing to allow 'lawless behaviour' to take place between the crew and the female slaves. In 1783 the captain of the *Zong* was short of water, so he threw 132 slaves overboard. The owners then brought an action for insurance alleging that the loss of the slaves fell within the clause of the policy, which insured against 'perils of the sea'. Chief Justice Mansfield, presiding over the case, said the slaves' situation 'was the same as if horses had been thrown overboard' and damages of thirty pounds were awarded per slave. The idea that the captain should be prosecuted for mass homicide never entered his head.

Naturally, the Africans frequently rebelled against such treatment. Insurrections were usually stillborn. the ringleaders were either killed or severely flogged. But the newspaper *Felix Farley's Bristol Journal* published a letter on 31 March 1753 by John Harris describing a massacre aboard the ship the *Marlborough*. Slaves from the Gold Coast rose up and killed many of the crew. The captain was wounded and climbed to the top of the mast. Harris was shot in the arm and stomach, but tried to pass it off lightly so that he would not be thrown overboard with the rest of the wounded. When they were almost back at the shore, they loaded up a longboat with goods. At this point, another set of slaves, also captive on the ship, but from a different tribe, used arms against the Gold Coast gang and forcibly boarded the longboat, sinking it. The Gold Coast slaves initially attempted to prevent them getting back on to the ship and a hundred died. The survivors fought the Gold Coast gang for a day and a night. As they drifted towards the coast, the slaves from the second tribe ordered a punt to be lowered into the water so they could reach the shore. Harris continued:

> For Fear of Delay, they thought proper to send a White Man in her; I was ordered in and out the Boat two different Times, when they sent two Blacks in stead, but the Cook told them, it was better to send me in the Boat, because I knew what to do in her. We loaded the Boat and rowed for our Lives to the Shore.

Harris then managed to find another slave trader to stow away in. The Gold Coast slaves still on the *Marlborough* cut the anchor rope and set sail. The last Harris saw of the ship was the captain being disembowelled and thrown overboard.

Once the slaves arrived in the West Indies, they were sold. Equiano was witness to this practice when he was sold himself:

> On a signal given (as the beat of a drum), the buyers rush at once into the yard where the slaves are confined, and make a choice of that parcel they like best. The noise and the clamour with which this is attended, and the eagerness visible in the countenances of the buyers, serve not a little to increase the apprehensions of the terrified Africans, who may well be supposed to consider them as the ministers of that destruction to which they think themselves devoted. In this manner, without scruple, are relations and friends separated most of them never to see each other again.

Buyers treated people from different tribes as if they were varieties of cattle or horses: Cormantis, for example, were thought to be proud and strong, but liable to revolt. There was a deliberate policy to buy a mixture of slaves from different parts of Africa so that they could not communicate with each other.

Those who were not sold were disposed of in what was called, for obvious reasons, a 'scramble'. At an agreed signal, the buyers would rush in and hang cards with their names around the necks of slaves. In this way they could buy 'job lots' cheaply. Sometimes sick slaves were bought as a speculation – they could be sold at a higher price if they recovered – but generally those who were ill were left to die. The result was that another 4.5 per cent mortality took place in the markets. Once purchased, the slaves were branded with their new master's initials using a red-hot iron.

It was into this atmosphere that a distant relation of Azariah Pinney – who had been transported to St Kitts and had then purchased a plantation on Nevis – arrived from Dorset. Azariah had been the despair of his father, John; Azariah's son, also called John, proved to be equally as much of a trial to his father. John never married and died very soon after Azariah in 1720. His cousins from Dorset inherited the plantation and, as they too were childless, sought an heir. A young cousin, the great-grandson of old John Pinney's daughter, Mary Clarke, had a granddaughter, Alicia Clarke, who had married Michael Prector. He described himself as a gentleman, though his detractors thought him 'worse than a footman'. He died early and left a little boy, John. The cousins took charge of him when he was fifteen, but did not treat him as an equal. His biographer, Richard Pares, in A West-India Fortune, described him as 'a somewhat underbred, obsequious apprentice with a little property of his own and much greater, though covert expectations'. John Prector's expectations were realised when his cousins died unexpectedly and left him Azariah Pinney's sugar plantation on the condition that he change his name to John Pinney.

Azariah Pinney had £15 when he was sent to the West Indies; he amassed a fortune worth £23,000, and the Lowland sugar-cane plantation in Nevis, but his son left debts of £6,800. This, at the age of 23 in 1763, was John Pinney's inheritance. A painting of him at the time shows a young man wearing an elegant frock coat and a floral waistcoat over a white shirt with ruffles down the front and at his cuffs, and a high, starched collar. His hair is scraped back and curled crisply about his ears, as was the fashion at the time. He stares warily, sharply, his mouth set grimly.

On his arrival, he was appalled to see men, women and children for sale – but it did not take him long to overcome his initial disgust. He wrote to a friend:

I can assure you I was shock'd at the first appearance of human flesh exposed to sale. But surely God ordained 'em for the use and benefit of us: otherwise his Divine Will would have been made manifest by some particular sign or token.

Pinney was to amass a fortune at the expense of enslaved Africans. He was perhaps typical of many plantation owners at the time; though not as cruel as some, he believed wholeheartedly in slavery. Although he was not outstandingly cruel to the slaves, he treated them with the same brutish disregard as most plantation owners. In his standing instructions to managers, he wrote:

> It is unnecessary, I flatter myself, to say a word respecting the care of my slaves and stock – your own good sense must tell you they are the sinews of a plantation and must claim your particular care and attention. Humanity tempered with justice towards the former must ever be exercised, and when sick I am satisfied they will experience every kindness from you, they surely deserve it, being the very means of our support.

Without a break in the paragraph he added, 'The latter must be kept free from ticks.'

Pinney was an extraordinarily precise man. He was neither particularly creative nor original but he possessed an inexhaustible fund of nervous energy and an overwhelming horror of debt. It is through his accounts that we understand his character, for he frequently annotated them. As Pares commented, 'It is hardly too much to say that John Pinney made a religion of his accounts; certainly he expressed his most passionate feelings in them, in long vehement comments on particular items, or in marginal rubrics.' When he later married and had children, he opened an account for each one of them at birth, charging them with the cost of their midwife and their share of the nurse he'd hired. He used the mortgage from one particular property in Dorset to pay for their upkeep. He once gave a present to a female slave: he sold her soap at a reduced rate. She was allowed to sell the soap for a profit, but he exacted his repayment to the last penny. When his brother-in-law became manager of Lowland, Pinney insisted that, because he was a qualified doctor, he should tend to the slaves when they were sick, but stipulated that he was not allowed to charge for this extra duty. He did, however, increase his manager's salary.

With the same chilling precision he applied to the lives of his children, he planned ahead for the slaves' food. They were given a plot of land in which to grow their own, and half a day off a week to tend it.

The plot was worthless to him and probably impossible for the slaves for it was too infertile to grow sugar cane. He expressly told his managers to keep altering when the slaves had their day off and sometimes cancel it altogether, so that they did not see it as a right. He instructed his managers to buy American corn in harvest time when it was cheap, and salt herrings from Glasgow to supplement their fare. The amount of corn he bought only lasted 35 weeks of the year, and the potatoes the slaves grew hardly covered the shortfall. Yet one year he castigated his manager for buying extra corn; another year he cancelled the pickled herring and ordered salt from St Kitt's salt pond instead. In addition they were given approximately eight pints of rum a week each, but only five during harvest time, the argument being that they could suck the sugar cane and drink the juice instead.

Ligon describes how, even though the Africans liked to eat corn on the cob roasted over a fire, they were given it as 'Loblollie'. The corn was stripped from the ear, pounded in a mortar and boiled in water before being served to them cold. He wrote, 'But the Negroes, when they come to be fed with this, are much discontented, and cry out, O! O! no more Lob-lob.'

Pares said:

Many colonies made no laws at all about the feeding of slaves before the humanitarians forced them into it at the end of the eighteenth century; and even where there were laws, the standards which they enforced were pitiably low. The French *code noir* stipulated for a supply of protein which would amount to little more than a kipper a day; and this code was not at all well observed. Some planters normally gave their slaves no food at all, but fobbed them off with payments or rum wherewith to buy food, or with Saturdays and Sundays to till their own provision grounds and feed themselves. The rum was drunk, the Saturdays and Sundays encroached upon or wasted, and the slaves starved. Their masters almost wholly disregarded their needs for protein, and could not see why they went on hunger-strike, or lost their sleep catching land crabs, or died. When I think of the colossal banquets of the Barbados planters, as Ligon describes them, of the money which the West Indians at home poured out upon . . . private orchestras and . . . Fonthill Abbey or even the Codrington Library, and remember that the money was got by working African slaves twelve hours a day on such a diet, I can only feel anger and shame.

The plantation owners often feasted until they were sick and frequently drank themselves stupid. Pages of Ligon's narrative describe the wealth of food the estate owners ate. A meal at the estate of Colonel James Drax, who was one of the first sugar-cane plantation owners, consisted of pork, chickens, goat, 'a Kid with pudding in his belly, a sucking pig, which is there the fattest, whitest, and sweetest in the world, with the poynant-sauce of the Brains, Salt, Sage and Nutmeg, done with Claret-wine', mutton, veal in a sauce of oranges, lemon and limes, turkeys, capons, hens, ducklings, turtle doves, rabbit, followed by oysters, caviare, anchovies and olives. For dessert there were custards, creams, cheesecakes and pastry puffs with fruit, including prickly pears and pineapples, served with twelve different sorts of alcoholic beverages.

The slaves lived in rough huts made of wood and mud, thatched with sugar-cane leaves. The floor was dirt and their beds were stuffed hessian sacks. At the beginning of the nineteenth century houses hewn out of coral were allotted to them in Barbados; none were larger than three by two and a half metres yet a number of slaves were expected to share one of these stone huts. They had an open fireplace and a few broken gourds to cook with. Once a year they were given a new outfit made of coarse linen or baize. The contrast with the 'Big House' could not have been more stark. A typical two-storey Caribbean plantation house was elegant, spacious and cool. Generally they were built of wood with shutters and hoods over the windows, like mini-roofs, to prevent the rain from blowing in. A long drive bordered with frangipani, hibiscus and bougainvillaea wound up to the veranda at the front of the house. Two thin rooms ran down each side: the ladies' sun-lounge, where they sewed and drank tea, and the master's study. The central portion of the house contained the living room and the dining room. The floors were polished pitch pine and the furniture was made from mahogany.

To make the slaves work eighteen-hour days they were often flogged. Pinney, in spite of his professed humanity, was not above instructing his slaves to be whipped, but, when Josiah Wedgwood, the potter who campaigned for the abolition of slavery, stayed at Pinney's house, Mountravers, Pinney wrote to his manager:

Do not suffer a negro to be corrected in his presence, or so near for him to hear the whip – and if you could allowance [i.e. give them extra rations of rum] the gang at the lower work, during his residence at the house, it would be advisable – point out the comforts the negroes enjoy beyond our poor in this country, drawing a comparison between the climates – show him the property they possess in goats, hogs and poultry, and their negro-ground. By this means he will leave the island possessed with favourable sentiments.

The type and kinds of punishments inflicted on the slaves were insanely cruel. Sir Hans Sloane collected specimens for the British Museum and was physician to the Governor, the Duke of Albermarle, in 1688 in the West Indies. He described in detail some of the tortures they had to endure:

The Punishments for Crimes of Slaves are usually for Rebellions burning them, by nailing them down on the ground with crooked Sticks on every Limb, and then applying the fire by degrees from the Feet and Hands, burning them gradually up to the Head, whereby their pains are extravagant. For Crimes of a less nature Gelding, or chopping off half of the foot with an Axe. These Punishments are suffered by them with great constancy ... For running away they put Iron Rings of great weight on their Ankles or Poltocks about their Necks, which are Iron Rings with two long spikes riveted to them, or a spur in the mouth ... For Negligence, they are usually whipt by the overseers with Lance-wood Switches, till they be bloody, and several of the Switches broken, being first tied up by their Hands in the Mill House ... After they are whip'd till they are Raw, some put on their skins Pepper and Salt to make them smart, and use several very exquisite Torments. These Punishments are sometimes merited by the Blacks, who are a very perverse Generation of People, and though they appear harsh, yet are scarce equal to some of their Crimes.

Understandably, there was a high slave mortality. As paying for a doctor was expensive, Pinney refused to allow his slaves to be treated when his physician brother-in-law left. He did fit up a former sugar-boiling house as a hospital, but it was near the manager's office so he could inspect them once a day. Pinney said they would merely malinger if left in their own huts. He dosed them himself, along with the help of his driver, Wiltshire, who ground roots to make compounds for treating venereal disease. Pinney's 1794 standing instruction to managers read:

> From experience I know it is better to depend on the adminis-tration of the simples of the country and with the assistance of Dr Buchan [author of *Domestic Medicine*], than to rest your dependence upon medical gent. in the island, who have estate pills and purges ready made up – probably from old medicines. When I took upon myself the care of my Negroes . . . I found my deaths not much more than one third, what they were before.

When Pinney arrived at Lowland many of the slaves were old and infirm and he set them free. While this sounds generous, it meant that he would not have to support slaves who needed medical care or whose work rate was low, nor did he have to pay taxes for them; these men and women would not have been able to find other work or accommodation on the island. He continued this practice, buying in 'man-boys' and 'women-girls' so that he would have a fit, young work-force. After thirty years, a third of his slaves, both the ones he inherited and the ones he had bought, had died.

The primary work for most of the slaves in the West Indies was cutting down sugar cane and manufacturing sugar. Work in the fields was gruelling. From January through to May cane was simultaneously planted, cut, milled and boiled. A drought at the start of the harvest could reduce the sugar content of the cane; heavy rain in spring might rot the cane in the ground. Bad land practices acquired in the colonies resulted in a decline in fertility; slaves had to fertilise the land using manure, seaweed and cane stubble. Erosion became a problem, which meant that the slaves had to dig depressions in the soil and pack the holes with manure to prevent rapid runoff of water. The work was laborious and humiliating: horse-drawn carts could not ride over the square depressions so the slaves had to carry animal dung across the fields in baskets balanced on their heads, dribbling liquid excrement over their faces.

Once harvested, the sugar cane had to be processed immediately to prevent the sugar in the plants' cells from going off. Two innovations characterised this period. One was the three-roller mill, which enabled plantation owners to increase their slaves' productivity from one-fifth of a ton of sugar a year each to half a ton. Again it was religion that helped sugar on its way: the Jesuits brought this new technology with them from China. The three-roller mill had the advantage that it could be adapted to animal, wind or water power, could be made in different sizes and needed only four to ten slaves to operate it. In Barbados there were over five hundred windmills; when there was a good, strong wind, of eleven to sixteen kilometres an hour, they could grind four to five tons of cane per hour. Once the juice had been extracted in the mill, it had to be taken to the boiling house almost immediately, as after a few hours it would start to ferment and would not crystallise. Sometimes the mill had to close temporarily to allow the boiling house to catch up.

The second innovation was the creation of the battery cauldron or the Jamaica train, which, in spite of its name, was invented on Barbados. Instead of boiling the juice in one cauldron, a whole set of them were used. Up to six at a time were fed from a central heating flue. The sugar-cane juice was heated in the first two cauldrons to allow impurities to rise to the surface so that they could be skimmed off. In the last three cauldrons the juice was boiled and ash or bone added to precipitate more of the impurities out, then boiled again until only half the amount remained.

About a tenth of the workforce laboured in the mill and the boiling house; it was a difficult and dangerous place to work. The slaves had to constantly stir the viscous syrup to prevent it burning and turning into caramel. In the final cauldron the syrup was reduced to magma – a thick paste saturated with sugar crystals. It was removed when the sugar master thought it had reached 'strike point' – when the sugar crystals and molasses were about to separate. During harvest, the mill and boiling house operated continuously with slaves working two shifts a day, partly due to the mistaken idea that sugar should not stop boiling until it 'struck'.

Barbadian colonist Thomas Tryon wrote:

In short 'tis to live in a perpetual Noise and Hurry, and the only way to render a person Angry, and tyrannical too; since the climate is so hot, and the labor so constant, that the Servants [slaves] night and day stand in great Boyling Houses, where there are Six or Seven large Coppers or Furnaces kept perpetually Boyling; and from which with heavy Ladles and Scummers they Skim off the excrementitious parts of the canes, till it comes to its perfection and cleaness, while other as Stoakers, Broil as it were, alive, in managing the Fires; and one part is constantly at the Mill, to supply it with Canes, night and day, during the whole Season of making Sugar, which is about six Months of the year.

The heat and noise were overpowering; the slaves risked losing fingers, arms or being badly burned in the mill and the boiling house. William Mathieson, nineteenth-century merchant and abolitionist, commented that the manufacture of sugar was the most onerous industry in the West Indies:

So rapid was the motion of the mill, and so rapid also the combustion of the dried canes or 'trash' used as fuel in the boiling house that the work of the millers and firemen, though light enough in itself, was exhausting. A French writer described as 'prodigious' the galloping of the mules attached to the sweeps of the mill; but 'still more surprising' in his opinion was the ceaseless celerity with which the firemen kept up a full blaze of cane-trash. Those who fed the mill were liable, especially when tired or half-asleep, to have their fingers caught between the rollers. A hatchet was kept in readiness to sever the arm, which in such cases was always drawn in; and this no doubt explains the number of maimed watchmen. The negroes employed as boilermen had a less exacting, but a heavier task. Standing barefoot for hours on the stones or hard ground and without seats for their intermissions of duty, they frequently developed 'disorders of the legs'. The ladle suspended on a pole which transferred the sugar from one cauldron to another was 'in itself particularly heavy'; and as the strainers were placed at a considerable height above the cauldrons, it had to be raised as well as swung.

Once the sugar had reached strike point, the final stage took place in curing or purging houses. The syrup was poured into earthenware cones, each one of which could hold twelve kilos of sugar. The cones were suspended in racks, pointing downwards; molasses drained from a hole in the apex of the cone leaving sugar crystals behind. This sugar was known as muscovado. It was ready after it had drained for a month and could then be shipped, often still oozing molasses, across the Atlantic.

Clayed sugar was of a higher quality. Wet caps of clay were placed on the top of the cones. As the water from the clay percolated through the sugar, it washed the molasses out. The sugar formed loaves and could be dried in the sun. The sugar that had been nearest the clay was white, but was brown in the middle and muscovado at the base of the loaf. Sir Hans Sloane recounts a story that sugar claying began when it was noticed that a hen, after foraging in wet clay, walked across wet sugar and left it whiter in those places where she had walked.

A yet more refined sugar was created by boiling and clarifying the sugar at least once more before straining it through a cloth and using the best clay to cap the cone. Most of this type of sugar was reserved for the planters. The Barbadians were also the first Westerners to use bagasse – the waste left over once the sugar cane had been crushed – to fuel the boiling house.

It was these combinations – sheer brute force and hard labour with skilled artisanal knowledge, and low-tech farming practices joined with high-tech refining – that turned the plantation system into a synthesis of field and factory. Mintz writes: 'The specialisation of skill and jobs, and division of labour by age, gender and condition into crews, shifts, and "gangs", together with stress on punctuality and discipline, are features associated more with industry than with agriculture – at least in the sixteenth century.'

John Pinney profited considerably from his plantation, Lowland, and from another he acquired, Gingerland. There is a strong possibility that he gained this second estate using rather underhand methods. Gingerland belonged to his neighbour, Roger Pemberton. As Pemberton was short of money, Pinney lent him large sums at a high rate of interest – 8 per cent, but Pinney had himself borrowed the money to lend to Pemberton from his friend, Thomas Lucas, at a rate of 5 per cent. Pinney denied that he was deliberately attempting to push Pemberton into debt, but he wrote to Lucas saying, 'should the worse come, I can only take possession of his estate, which is so advantageously situated to mine, that I can work it at half the expense it is now worked, nay, it will make me a complete estate from sea to mountain and it will make an addition of 40–50hhds [hogsheads; each hogshead was somewhere between 660 and 813 kilos (1,456 and 1,792 pounds)] of sugar per annum . . . If that estate comes into my possession, it will make mine the completest and (I think) best single estate in Nevis.' Eventually, Pemberton did have to let the estate go and, after leasing it to another tenant, Pinney became the sole owner. He wrote, 'From that estate which came to me heavily burdened with debt and an annuity, after purchasing upwards of 100 negroes, the estate, erecting new buildings and a windmill, I made my fortune.'

Pinney remained on his plantation for longer than many other owners. Increasingly plantation owners returned to Britain, leaving their land in the hands of overseers, who were arguably even crueller to the slaves. Pinney, too, had always wanted to return to England eventually, and he was concerned that his younger children, being educated back home as the sons and daughters of a wealthy man, were being unbearably spoiled. It was his oldest son, John Frederick, reading law at home and disinclined to live in Nevis, who really pushed his father to sell. It made sense to offer the plantation to a neighbour, J.H. Clarke, but another man, Edward Huggins, was also interested. Pinney asked his friend, James Tobin, to act on his behalf but carried on regardless himself, selling his estate to Huggins at the same time as Tobin sold it to Clarke. Huggins retained the estate in 1808, but Pinney and Tobin fell out for some time over this debacle.

Pinney clearly had a choice over which man to sell to, yet he sold his plantation and slaves to a man renowned for his cruelty. The record is unclear, but there were between five and nine attempts on Huggins's life by his own slaves. When Pinney's slaves heard that Huggins had bought the place, they ran away. Huggins put his son, Peter, in charge as a manager. There were a number of unpleasant scenes: in one instance, Peter tried to show a slave how to plant sections of cane, but the slave persisted in using the method Pinney had taught him and was rude to his manager. Peter seized hold of the man, whereupon the whole slave gang downed tools and advanced. Peter screamed that he would put to death the first one that reached him, and the slaves retreated. The unruly slave was punished and the rest of the workforce made to work at night as well as during the day (an illegal punishment by this time).

Presumably as a result of these difficulties, on 23 January 1810 Edward Huggins took Pinney's slaves to the marketplace where they were publicly whipped for hours. He ordered the men to be flogged between 47 and 242 times, and the women between 56 and 212 times. Even though three magistrates were present, no one stopped him. It is hardly likely they would have done: one magistrate was his own son, Edward. One other was a doctor, and a member of the clergy was also present, yet the only one who voiced a complaint was the doctor – when he felt that one particular slave had had as much as he could stand. No one asked what the slaves were alleged to have done. One of the women died. Some said Huggins had authorised them to be whipped lightly so that more did not die; others said black people didn't feel pain in the same way as whites.

Only Tobin wrote to the governor to complain about Huggins's cruelty. Although Huggins was tried, he was acquitted. Furious, Tobin alleged that the judge, prosecutor and jury were in Huggins's interest if not pay. He accused the Chief Justice of St Kitts of pronouncing sentence under the influence of liquor upon a case he had not heard and the Chief Justice of Nevis of threatening at a council meeting to shed Tobin's blood. The Hugginses didn't stop there: in 1812 Edward Huggins shot a black man dead and in 1817 the older Huggins presided over another terrible flogging. The Governor of the Leeward Islands, a man named Elliott, wrote to the House of the Assembly for the British Caribbean Colonies about the case, saying that justice could no longer be carried out in the islands, which won him nothing but vicious contempt from the planters.

As for Huggins, when Pinney heard of the flogging, he wrote to Peter, but his letter is hardly harsh:

I am truly sorry to hear that the negroes have not behaved as they ought and you wish. During my long residence in the island no people could have behaved better in general than they did – it will give me infinite pleasure to hear that all things go on to your satisfaction and that the negroes are reconciled to their change of masters. We know the disposition of these people that they are apt to try the temper of a new master by not being as correct in their conduct as they ought at first, but I have no doubt with kindness and attention to their little wants, when they deserve it you will find them a well disposed set.

While Pinney had been abroad, Bristol had grown to become the largest slave-trading port in the country and it was in Bristol that Pinney settled. Using the money he made from the sale of his sugar plantations, he increased his fortune fivefold.

His study overlooked the street, a dark, elegantly furnished room with heavy mahogany shelves and a desk, but it was the top floor of his Georgian house, just off Park Street in Bristol, that must have held the most fascinating view. From here Pinney would have been able to look down upon the River Severn, which flowed right into the heart of the city, boats docking and unloading literally between the streets (the river has since been rerouted). The poet Alexander Pope wrote of it in 1739, 'In the middle of the street, as far as you can see, hundreds of Ships, their masts as thick as they can stand by one another, which is the oddest and most surprising sight imaginable . . . a Long street full of ships in the Middle and Houses on both sides looks like a Dream.'

Below Pinney was a mews street where he had the offices for the company he formed with Tobin, with whom he had reconciled. The two worked on behalf of plantation owners, selling sugar, providing ships to transport their sugar and, above all, financing the planters' extravagant lifestyles at a profitable rate of interest. Even though he'd sold his own plantations, he acquired a further seven when the owners could no longer repay their debts. At his death in 1818, he left £340,000, worth £17 million today.

What fuelled the slave trade was not just the plantation owners' desire for slaves: the trade was triangular – British goods were bartered in Africa; Africans were sold into slavery destined for the West Indies, and sugar was exported back to Britain from the colonies. Later a second triangle developed that was equally cruel but far more bizarre: New England (an English colony in America covering what is now Maine, New Hampshire, Vermont, Massachusetts, Rhode Island and Connecticut) sold rum to Africa in exchange for slaves that were sent to the West Indies, then shipped the product of the very sugar cane these slaves lost their lives for – molasses – back to New England to make more rum.

England only fully entered the slave trade in 1660 when Charles II founded a new company, the Royal Adventurers into Africa, which was granted a monopoly over the English slave trade for a thousand years. Its financiers were the backbone of England: they included seven knights of the realm, four barons, five earls, a marquis, two dukes and four members of the royal family. The first ships took slaves from the African gold coast in Guinea to Surinam and Barbados. Within two years, their annual return was £1 million.

Although the company was closed in 1672, it was relaunched, with only minor changes in staff and shareholders, as the Royal African Company. Its credibility was guaranteed by the king himself: his warrant on 27 September 1672 read, 'We hereby for us, our heirs and successors grant unto the same Royal African Company of England . . . that it shall and may be lawful to . . . set to sea such as many ships, pinnaces and barks as shall be thought fitting . . . for the buying, selling, bartering and exchanging of, for or with any gold, silver, Negroes, Slaves, goods, wares and manufacturers.'

By 1689 the company had transported 90,000 slaves. Its profits continued to soar and many well-known people became shareholders: poets Alexander Pope and John Gay; the king's mistress, the Duchess of Kendal; and Sir Thomas Guy, the bookseller and philanthropist, who, when he died, bequeathed money to found a hospital in London for the 'poorest and sickest'. Guy's Hospital has since grown to become one of the most famous in the world.

However, the company's supremacy was challenged long before they lost their monopoly. In 1676 the British Government heard from a Royal African Company official that:

Several ships have lately arrived, at our Island of Barbados, from those parts of Africa with Negros and other goods and several others are now on the said coast; all of which are set out by private Traders. And that the said African companys agents seising 80 Negros, part of 150 negros so imported . . . in the Ship Providence; the same were violently taken away from them, and they are those who assisted them, beaten and wounded.

Speightstown, in Barbados, even became known as Little Bristol because of the preponderance of Bristol ships trading there. This is the only evidence that exists to prove that Bristol was involved in an illegal trade before 1690, partly because, according to Dr Madge Dresser, a historian and author of *Slavery Obscured*, Bristolian ships commonly stated that their destination was Madeira or the Cape Verde islands when they were actually travelling to Africa. Current records show that three sailed for these professed destinations in 1679. One was called the *Blackamore*, which is in itself suggestive; in addition, it was provisioned by John Cary, a merchant who was very much in favour of slavery and who drew up a petition to Parliament to let the merchants of Bristol share in the African trade, and by two sugar merchants, Thomas Deane and James Pope.

John Cary, in 'A Discourse of the Advantage of the African Trade to his Nation', wrote:

> The African trade is a Trade of the most Advantage to this Kingdom of any we drive, and as it were all Profit, the first Cost being little more than small Matters of our own Manufacturers, for which have in Return, Gold, Teeth [ivory], Wax and Negroes, the last whereof are much better than the first, being indeed the best Traffick the Kingdom hath.

The Company finally lost its monopoly in 1698 and Bristol quickly became officially the greatest slave-trading port in England. The city was uniquely well suited for this pre-emininence – merchants were practised at fitting out ships, for the port had long run a lucrative trade in kidnapping and transporting indentured servants and had experience of sailing down the coasts of France and Spain to Africa, where they had bought copper from the Africans. This had taught them that the Africans would not buy shoddy goods. As S.I. Martin points out in his book *Bristol's Slave Trade*, this exchange was not between civilised men and savages, rather it was between English merchants and captains and African kings, chiefs and dignitaries.

Originally textiles were the largest element of merchandise loaded on the Guinea-bound ships: baffts, brawles and romallos from India, serges and perpets from Devon, and striped and coloured cottons from Lancashire. Brandy and gin cordial were included to lubricate the negotiations. Goods made in Bristol consisted of glassware, umbrellas, 'negro hats' edged in gold, silver or copper, as well as more traditional metalwork: brass pots, brass bangles or manillas (iron and copper rods that served as currency), guns, cutlasses, knives and gunpowder. By 1728 over 21,000 small arms had been shipped to West Africa from Bristol gunsmiths via Bristol ships.

Originally a slave trade had operated between the Arabs and the Africans, with the Arabs supplying brass, and Bristol was fortunate because they discovered they could mine the necessary minerals for making brass from African copper in the hills around the city. Within a few years, dozens of small metalworkers had sprung up along the River Avon. As a result Bristol became an intimate part of both the sugar and the slave trade, as dozens of small refineries were built within the city centre.

As you take the train into Bristol today, the first things that strike you are the rows of multicoloured houses, bright as sweets, that arch along the skyline. One of these streets runs just above the docks, from St Mary Redcliffe Church towards the Ostrich Wharf. The houses on Redcliffe Parade are mustard, peppermint, terracotta, blackbird-egg blue and mint. In one of them lived a ship owner, Thomas King. In the eighteenth century he traded directly with Africa, and from his house he could watch his own ships, including the *African Queen*, being built.

Opposite his house is the Hole in the Wall pub. It used to be called the Coach and Horses and is distinguished from any other pub by a tiny spy house that looks out on to the docks. These were the days when men reluctant to work on slave ships were kidnapped and forced to become sailors: the spy house was a lookout for these press gangs. It was no defence against a more insidious form of control: landlords frequently took money from ship owners in order to ply sailors with drink so they would slide into debt. Unable to pay, the only way they could avoid prison or the poor house was to enlist on a slaver.

The Hole in the Wall is on the corner of Queen's Square, a beautifully elegant square completed in 1727 when Bristol's slave trade was at its height. It was built using the wealth generated by sugar and the slave trade, as was much of the city centre. Round the corner from the square is an almshouse for old sailors founded by the Society of Merchant Venturers, now painted the colour of strawberry ice cream, its tiny courtyard choked with pot plants and plastic windmills. The Severn would have flowed a few metres away from the almshouse; on the other side was Lewin's Mead Sugar House where the raw cane sugar from Barbados was unloaded on to the quay directly in front of it and refined. It has been artistically converted to a Hotel du Vin that ironically specialises in serving different varieties of rum. Almost next door is another pub, the Three Sugar Loaves. It was built on the site of a sugar refinery destroyed by fire in 1859; its sign is a conical sugar loaf and pair of sugar tongs used to handle the lumps of sugar when the loaf was broken apart.

At least eight families lived here whose livelihoods revolved around these twin trades of slavery and sugar; in fact, very few streets or buildings in the city have no connection to them. Even Bristol's patron, Sir Edward Colston, who founded the Colston Hall (mainly used for musical performances), two schools, a hospital, three charities and an almshouse, was an official in the Royal African Company, and the minutes from 1698 show that he was present at meetings where he organised and approved the sale and transport of Africans to the Caribbean. His almshouse is panelled using wood from a slave ship; a painting showing him on his deathbed features a female African slave kneeling beside him. However, this picture was painted a hundred years after he died; the artist had his body disinterred so that he could depict Colston's true likeness. It is thus not known whether he really owned a black slave or whether she is a metaphor in oils.

Of course, Bristol is not the only town founded on the back of sugar and the African slave trade. Liverpool usurped Bristol's position as the largest slave port by 1770; the depth of the Mersey meant that larger boats could dock in the city and slaving could continue more economically. More than 4,500 slavers trading in a million slaves were to leave Liverpool. The actor George Cooke (1756–1812) once performed in Liverpool's Theatre Royal. As he was somewhat drunk, he was booed by the crowd. He shouted back, 'I have not come here to be insulted by a set of wretches, every brick of whose infernal town is cemented with an African's blood.' To their credit, the audience applauded him.

Today, it is not just parts of Liverpool, London and Bristol that exist due to these trades: the new nobility who had made their money from sugar contributed to the purchase and erection of paintings, museums, art galleries, theatres, libraries and banks throughout the country, all of which we use today, unthinkingly and unknowingly. The true cost of these luxuries was staggering. Between 1700 and 1800, 263,000 slaves were imported to Barbados alone. In 1500, the population of the African continent was estimated at 47 million. Over the next 350 years between ten and fifteen million Africans were shipped to the New World. Another four to six million died en route. This is a total of between fourteen and twenty-one million people sold into slavery, many of them to plant, hoe, harvest and refine sugar from sugar cane.

4. CREAMS, CAKES, CUSTARDS AND CHARLOTTES

To make Conserve of red Roses, or any other Flowers
Take rose-buds, or any other flowers, and pick them; cut off the white part from the red, and sift the red part of the flowers through a sieve to take out the seeds; then weigh them, and to every pound of flowers take two pounds and a half of loaf-sugar; beat the flowers pretty fine in a stone mortar, then by degrees put the sugar to them, and beat it very well, till it is incorporated together; then put it into gallipots [a small earthenware pot], tie it over with paper, over that a leather, and it will keep seven years.
Hannah Glasse, *The Art of Cookery Made Plain and Easy*, 1747

What fuelled the ignominious scramble for sugar colonies, and the concomitant rise of the slave trade, was the ever-increasing desire for sugar in the mother countries. Sugar was first used in England in the thirteenth century as a medicine, just as it had been initially in India. However, it now became a drug of the wealthy. A German traveller who met Elizabeth I in the sixteenth century commented on England's love of sugar:

The Queen, in the sixty-fifth year of her age, very majestic; her face oblong, fair but wrinkled; her eyes small, yet black and pleasant; her nose a little hooked, her lips narrow, and her teeth black (a defect the English seem subject to, from their too great use of sugar).

During meals the nobility served 'subtleties' – monstrous concoctions made of confectionery. Robert May, a professional cook who lived during the reigns of Elizabeth I, James I, Charles I, Cromwell and Charles II, created recipe suggestions for aristocrats who wished to copy royalty. He described how to make ships of pasteboard in case they could not afford to fashion the whole vessel from marzipan. He explained how to model a sugar sculpture of a stag that bleeds claret wine when an arrow is removed from its flank, a castle that fires artillery at a man-of-war and gilded sugar pies filled with live frogs and birds. At the end of the display, he encouraged ladies to toss eggshells of scented water at one another to subdue the smell of gunpowder released by the sugared fortress.

As well as being used decoratively, sugar was increasingly employed as a preservative and a spice. May recommended using different grades of sugar: fine white for macaroons, loaf sugar for marmalade, raspberry jelly and cakes, and to preserve cherries, he wrote, 'if you would have them pure-coloured do them with the best sort of sugar'. A recipe from 1693 for preserving flowers suggested, 'To Pickly cowslip flowers, lay a laying of flowers and a laying of sugar, till the jar or pot be full' and 'To make a syrup of violets, take the deepest and best coloured violets and to every pint of liquor put two and a half pounds of right Brazil sugar.'

Although sugar was treated as a spice, we are used to thinking of sugar *and* spice. Sugar changed from being treated as a luxurious addition to food, as if it were a spice, to a staple food, at least for the aristocracy, from the sixteenth century onwards as sugar became more plentiful. Sidney Mintz, in his book *Sweetness and Power*, points out that sugar as a spice now only survives in fringe areas in food traditionally associated with festivities – biscuits with ginger and cinnamon, fruit jellies served with duck or goose, in the thick, crisp brown sugar and clove crust on a ham, which, as he says, 'demonstrates what anthropologists have long contended – that holidays preserve what the everyday loses'. Sugar remained, though, a preserve of the wealthy.

Just as slavery was expanding and sugar production was increasing exponentially, a discovery was made that would one day lead to the demise of the sugar-cane industry. It was to become a tool that one religious group would use as a weapon in the war against slavery, and, two centuries later, it would create havoc for the world's poorest people. A German physicist discovered how to extract sugar from a European plant.

Beta vulgaris, the sugar beet, is a root vegetable originally from the Mediterranean. Sugar beet is part of a family that includes Swiss chard and beetroot. It was first mentioned in the eighth century BC in an inventory of the kitchen garden of a Babylonian king. References have been made to it in Greek and Roman classical literature, as it was initially cultivated for its medicinal properties: Hippocrates said beet broth could be used to treat certain ailments and that beet leaves dipped in wine would help wounds heal. It was thought to help nose and throat complaints and was considered a good hair tonic. It wasn't until the sixteenth century that people realised the plant stored sugar in its bulbous root. In the early seventeenth century a Frenchman, Olivier de Serres, wrote, 'The beet root when boiled yields a juice similar to syrup of sugar.'

Yet a process for extracting this sucrose was not developed until 1747. Andreas Marggraf, a professor of physics from Berlin, sliced, dried and pulverised beets. He added alcohol to the powder and brought the liquid to the boil. He then filtered it into a flagon, corked it and left it for some weeks. To his surprise, he discovered that sugar had crystallised around the inside of the flask. In further experiments, he found that he could extract fifty grams of sugar from half a kilo of dry beet, but the process was expensive and time-consuming and no one paid much attention to his research. It would be another half a century after Marggraf's original discovery before the sugar-beet industry was revived and would contribute to the death of the slave trade.

The cane that conquered the world, *Saccharum officinarum*, which became known as the Creole cane, was slender and dark compared to contemporary canes. It was taken from India and from there it had spread to the Mediterranean before being carried to every colony the West possessed, travelling under the guise of religious instruction, transported by armies, given as the gift of sultans and sheikhs to kings and chiefs, offered by missionaries, shipped by captains. It wasn't until 1768, almost ten thousand years after the domestication of the first cane, that a new variety was introduced.

The first navigator to sail round the world was a Frenchman, Louis Antoine de Bougainville. He found a new variety in Tahiti; a grasshopper-green cane that turns yellow as it matures. He took samples of it to Mauritius, then called the Ile de Bourbon; as a result the Otaheite cane was sometimes referred to as the Bourbon cane. De Bougainville even grew it in his own garden at Suynes near Paris, although it only survived there for three years. Captain Bligh carried the Otaheite cane to St Vincent on his second voyage to the island in 1793 (his first voyage had resulted in the infamous mutiny on the

Bounty); this initial shipment consequently became the parent stock for the French West Indies and ultimately the English West Indies.

Like carrying coals to Newcastle, the Otaheite was transported from Mauritius to India by Captain William Sleeman in 1827 and was planted first in the Government Botanical Garden at Calcutta and then at his estate at Jubbulpore. For this he received the Gold Medal of the Agricultural and Horticultural Society of India. Captain Sleeman had another claim to fame. He was single-handedly responsible for eradicating a secret society that worshipped the goddess Kali and killed travellers by strangulation. They called themselves Thugs; the name derives from the Hindi verb *thaglana*, 'to deceive', for they initially befriended travellers before dispatching them. It was estimated that 30–40,000 deaths every year were attributed to the Thugs.

Twenty years later Captain Cook, an officer serving in India, took the Otaheite cane from Captain Sleeman's estate to Peshawar where it remains to this day, growing along the Ganges as a garden vegetable. The naturalist Humboldt described it as 'one of the most important acquisitions for which colonial agriculture is indebted to the travels of naturalists. It yields not only one-third more juice, but from the thickness of its stem and the tenacity of its ligneous fibres it furnishes much more fuel.' As Humboldt realised, the Otaheite contained far more sucrose than the Creole and it is for this reason that the cane spread with amazing rapidity, almost completely replacing the Creole in Java, Burma, Mexico, the Philippines, Hawaii, Jamaica, the British Caribbean, the French West Indies, Cuba, Puerto Rico and British Guiana. For the best part of a century the planters destroyed the original Creole cane and Otaheite was the main cane grown.

It was the Jesuits who carried sugar cane to America. They had transported sugar cane to California in 1750 but they also introduced it to the southern coast of America, Louisiana. Though Louisiana had been sighted by the Spanish, it was, for most of its colonial history, owned by the French. The Jesuits, travelling with a ship carrying troops destined for St Domingue in 1751, stopped at Louisiana and sent African slaves and sugar cane to their branch house where Canal Street in New Orleans is now. In New Orleans they grew the cane in the gardens of their missionary centre, but they, like the richest planter of the time, Dubreuil, who obtained cane seedlings from the Jesuits, were unsuccessful. It was not an ideal habitat. As Jock Galloway says, 'An early frost in late November could damage unharvested cane while a late frost in March killed new cane as it appeared above ground. Severe winter frosts killed the cane in the ground.'

Louisiana remained French until 1762, then fell to Spanish rule, before being taken back by the French in 1803, finally becoming part of the United States of America a year later. It was during Spanish rule in the later eighteenth century that sugar-cane production really took root. A Spaniard, Solis, was growing cane at Terre-aux-Boeufs for rum making. He sold his plantation to Antoine Mendez, who called in the help of Marin, a technician who had been in St Domingue studying cane cultivation and manufacture. The two of them managed to make loaves of white sugar, which they produced at a banquet. The Mayor of New Orleans, Etienne Boré, was instantly captivated. Mendez supplied him with cane to cultivate his own plantation and he hired Marin to lay out a factory based on those he'd seen in St Domingue. This time the mission was successful and by 1796 Boré had established Louisiana's sugar industry.

The best land to grow sugar cane in Louisiana was along the flood plains of the Mississippi and in the bayous. Transportation of the cane to the mill was by water so each person's plot was narrow to allow every planter access to the river. As the area frequently flooded, the plantation owners' mansions were built above the levee. The result was a linear settlement pattern that has persisted to the present day.

Louisiana was so successful that up until the American Civil War Louisiana produced 95 per cent of sugar in the American South. As with all the other colonies, its productivity rested on its use of slaves and it was the last of the sugar colonies run on the plantation system. Less than fifty years after Boré's first sugar loaves were refined, the slave population had reached a massive quarter of a million, most of whom were working in sugar production. But sugar in America was to receive a triple blow: war, disease and the abolition of slavery would all but ruin the industry.

The first introduction of cane to Australia was in January 1788 when the First Fleet, a collection of convict ships, left England and arrived in Botany Bay carrying sugar cane. But no one knows what became of the sugar cane that was part of the cargo.

One of the first pioneers in the antipodean industry was Thomas Scott. His father owned the Golden Grove estate in Antigua where he'd been a manager. The London Missionary Society employed him for three years to grow sugar cane in New Zealand. On his arrival in Sydney in 1819 he was offered a salary of £250, a large sum at that time, by Colonial Secretary Major Goulburn to grow cane at Port Macquarie in New South Wales. Perhaps wisely, he turned this down and established his own sugar-cane fields in Prospect on the Hastings River at the mouth of which is now Port Macquarie. His first yield from 243 hectares was 71 tons, but he lost the lot to a fire. He was given land at Brisbane Waters, a few miles north of Sydney, and carried out experiments in planting and breeding. In 1863 he supplied Captain Louis Hope with 200,000 plants to establish sugar cane in Queensland. The first sugar from this region was allegedly refined in either Brook's biscuit factory or Fowles's bakery in Queen Street, Brisbane. These early manufactures made the princely amount of three kilos.

It took a Barbadian, John Buhôt, to properly kick-start the industry. Hope planted and milled the cane and Buhôt refined it in a factory built in 1863. Queensland also developed a floating mill, the *Walrus*, that operated along the Brisbane River. Although, in the 21st century, Australia can rival the major sugar exporters of the world and has one of the most advanced scientific institutions devoted to the study of sugar cane, sugar in Australia did not become nationally important until the early twentieth century.

Australia was the first colony that did not use slave labour, although Queensland, for instance, employed penal labour. Indentured labour from the Pacific islands, such as Polynesia, was imported to keep costs down. Although New South Wales did not subscribe to this practice, they hired Polynesians who had worked out their contract in Queensland. Galloway comments:

> In the new colonies of the Pacific, as centuries before around the Mediterranean, on the Atlantic islands and in colonial America, the demands of the sugar industry transformed landscapes, destroyed indigenous societies, triggered flows of immigrants and created a world of its own. It was, along with the western imperialism with which it was so closely linked, a major factor in the formation of the cultural geography of the tropical world.

So sugar had come full circle: after originating in New Guinea, it had travelled the world, before finally returning to the Pacific to be cultivated by the original race who had domesticated cane in the first place.

It seems fitting that around the same time that sugar cane spread almost to the opposite ends of the earth – to America and Australia – sugar changed from being a luxury only the wealthy could afford to an everyday feature of most Westerners' lives. From 1700 to 1800, sugar consumption increased by a massive 400 per cent in England – from two kilos per person to nine kilos (at the beginning of the 21st century it has now reached 35 kilos per person per year). Mintz writes:

> The steps by which England shifted from buying modest quantities of sugar from Mediterranean shippers; to importing ... a somewhat larger supply; to buying yet larger quantities from the Portuguese, first in the Atlantic islands and then in Brazil, but refined outside England; to establishing her own sugar colonies – first to feed herself and to vie with Portugal for customers and then, with time, simply to feed herself, finishing the processing in her own refineries – are complex, but they followed in so orderly a fashion as to seem almost inevitable. On the one hand, they represent an extension of empire outward, but on the other, they mark an absorption, a kind of swallowing up, of sugar consumption as a national habit. Like tea, sugar came to define English 'character'.

The woman who could be held responsible for introducing sugar to the middle classes and bridging the gap between the aristocratic use of sugar and the everyday consumption is Hannah Glasse. Her two books, *The Art of Cookery Made Plain and Easy* and *The Compleat Confectioner,* were published in 1747 and 1760 respectively. She described how subtleties could be further modified from Robert May's extravagant creations to appeal to a wider audience: how marzipan and blanched almonds could be crafted into a 'Hedge-Hog', for instance. She gives the recipe for ten different items: hedges, gravel walks, a little Chinese temple, the top, bottom and sides arrayed with 'fruits, nuts of all kinds, creams, jellies, syllabubs, biscuits, etc., etc.' Mintz says of her:

> Mrs Hannah Glasses's special confectionery cook book ... probably contributed to the behavioural bridging between matron

and drudge that accompanied the emergence of new middle-class segments. It offers good evidence of how comprehensively sugar was entering the English diet. This pathbreaking work dealt not only with sugar sculpture frames and mini-subtleties, but also with sweetened custards, pastries and creams, the recipes for which required port, madeira, sack [sweet sherry], eggs, cream, lemons, oranges, spices and immense quantities of sugar of many sorts. By instructing the rising middle classes in the fabrication of pastries and other desserts, Mrs Glasse provided rich documentation that sugar was no longer a medicine, a spice, or a plaything of the powerful – though of course the powerful would continue to play with sugar, in new ways.

Marie Antoinette may have instructed French peasants to eat cake but, as sugar became less expensive, it was the English middle class who responded. At first a pudding could be part of a second or third course, along with meat or fish, but by the end of the nineteenth century they began to follow the savoury courses as courses on their own. Elizabeth Ayrton, author of *The Cookery of England* in 1974, writes:

> In the first part of the eighteenth century a 'pudding' almost always meant a basis of flour and suet with dried fruit, sugar and eggs added. As the century went on, hundreds of variations were evolved, recipes multiplied; even the plainest dinner served above the poverty line was not complete without its pudding. Hot puddings, cold puddings, steamed puddings, baked puddings, pies, tarts, creams, moulds, charlottes and bettys, trifles and fools, syllabubs and tansys, junkets and ices, milk puddings, suet puddings: 'pudding' used as a generic term covers so many dishes traditional in English cookery that the mind reels as it dwells on these almost vanished splendours of our tables.

Although the French make beautiful desserts and gorgeous tarts, cheese has a firm place at the end of a meal. Sweetness is treated almost like a spice in France, as it has been in China throughout the history of sugar. Mintz says:

> The less conspicuous role of sugar in French and Chinese cuisines may have something to do with their excellence. It is

not necessarily a mischievous question to ask whether sugar damaged English cooking, or whether English cooking in the seventeenth century had more need of sugar than French.

Now a sweet dessert at the end of a meal has become a common feature of meals in much of Europe and America. Viewing our eating habits from an anthropologist's perspective, Mintz adds, 'There is nothing natural or inevitable about eating sweet food at every meal or about expecting a sweet course ... Yet it is by now so commonplace that we may have difficulty in imagining some completely different pattern.'

Sugar infiltrated even the working class through tea drinking. As tea and sugar became increasingly cheap, they were drunk together and could make a poor meal feel hot. Sweet tea and white bread became the standard meal for many of the working class, especially women and children who saved what meat they could afford for the men. Gradually, as the price dropped, more people bought sugar already incorporated into food – in jams, biscuits, tarts, buns and sweets. These baked goods and confectioneries grew to be a traditional accompaniment to a hot, sweet drink of tea or coffee. At the start of the nineteenth century 2 per cent of the average person's calorific intake came from sugar; a century later it had risen to 14 per cent. Today it is more than 20 per cent.

But there was a price to pay for our love of sugar and it was already beginning to be felt in the nineteenth century. Dental caries – tooth decay – does not usually exist in hunter-gatherer societies, as their diets are largely still devoid of refined sugars and carbohydrates. In England, in the second half of the nineteenth century, when sugar became widely available to every member of society for the first time, there was a sharp rise in tooth decay.

5. DANTE'S HELL OF HORRORS

> This untamed fire of justice continues to burn in the affairs of man, and it lights the way before us.
> George W. Bush, Speech on Goree Island, Senegal, 8 July 2003

It was June 1675 in Barbados. A female house slave, Fortuna, heard rumours that a band of Cormantis, slaves from the Gold Coast, were plotting a general uprising. The Cormantis had a reputation for sheer hard work, as well as being the most likely to cause trouble. In Edward Long's *History of Jamaica*, published in 1774, it was stated, 'Nature does not instruct the farmer to yoke tigers to his team or to plough with hyenas.' In the same book Christopher Codrington added, 'The Cormantis are not only the best and most useful of slaves, but are all really born heroes . . . No man deserved a Cormanti that would not treat him like a friend rather than as a slave.'

In 1675 the Cormantis planned to murder all the whites on the island apart from the prettiest women and install a Cormanti called Cuffee as king. Fortuna warned her master, a planter called Captain Giles Hall, two weeks before the uprising was due to take place. He in turn alerted the governor who asked a dozen officers to look into the allegations. Six slaves were burned alive; eleven others were beheaded and dragged through the streets of Speightstown. A total of 35 men were executed, and Fortuna was rewarded with her freedom. In Barbados further proposed uprisings ended equally brutally in 1683, 1686, 1692 and 1701.

The western world was growing to rely on sugar, but the slaves producing the sugar were increasingly becoming a problem for the white plantation managers. Barbados was as unstable as liquid gelignite. The island was so densely populated that there could be no halfway measure: any uprising would have to be a colony-wide conspiracy, as there was literally nowhere to hide and nowhere for the rebellious slaves to escape to. The white population had dwindled owing to a combination of absentee planters and the loss of indentured white servants, and the planters were living in a state of siege. Their idea of buying slaves from different ethnic groups had not worked as the slaves had been on the island long enough to develop a Creole patois. In addition, as sugar profits narrowed, planters were tempted to feed the slaves less and work them harder. Their crowning act of stupidity was to arm some of the blacks to bolster their own militia who were now better equipped to fight their masters.

The earliest recorded revolt took place in 1522 in Hispaniola and involved a hundred slaves, many of whom were slaughtered. Between 1640 and 1713 there were seven rebellions on English colonies. The island where the slaves were most successful was Jamaica. The escapees were known as maroons. The root of the word is unclear: it could be derived from the Spanish *marrano* (wild pig), or *simaran* (monkey) or from *cimarron* (mountain dweller). What we do know is that it led to our word marooned. The simple reason why slaves managed to escape in Jamaica was because there was somewhere apart from the interminable fields of sugar cane for them to hide – Jamaica has a forested, mountainous interior. Of course, maroons frequently stole from the plantations and they were a continual reminder to the remaining slaves that freedom was possible. Although they were hunted down like animals, at least some managed to survive, setting up a famous colony in Palmares that lasted from 1605 to 1694. The English were unable to subdue them and, after the First Maroon War in 1740, a peace treaty was signed, guaranteeing them freedom, the right to land and permission to sell their crops. Their legacy lives on in Jamaica; maroon villages on the outskirts of sugar-cane plantations have grown to become towns.

Richard Dunn, author of *Sugar and Slaves: The Rise of the Planter Class in the English West Indies 1624–1713*, wrote:

> The Jamaica slaves put the lie to the planters' contention that Africans were really happy in their bondage; on the only island where rebellion had a chance of success, it happened often. Yet even in Jamaica the blacks were rather ineffectual rebels, a key reason being that in nearly every conspiracy some loyal slave betrayed the secret to his master. Here is the supreme irony. The English planters, who treated their slaves with such contemptuous inhumanity, were rescued time and again from disaster by the compassionate generosity of the Negroes. In consequence the slave uprisings – even in Jamaica – caused less damage to the planters than hurricanes, earthquakes, malaria epidemics, and French raids.

St Domingue, the western part of Hispaniola, which the Spanish ceded to France in 1697, became a byword for profusion and extravagance among the planters. As Noël Deerr wrote, 'Within a few years a community of pirates, buccaneers, filibusters and smugglers had grown into a wealthy association of planters.' In 1718 a contemporary writer had this to say of the inhabitants:

A change of manners had come over the early pioneers. In place of bananas and a piece of wild pig on which they feast after having had the task of finding it in the forest, one sees on their tables relèves of game and of made-up dishes. The best wines of Burgundy and Champagne are not too dear for them, and, whatever is the price asked, they pay it.

The slaves on St Domingue were able to do more than exist on the margins of a so-called society, eking out a living and begging to sell their own food, and, in 1791, a victorious slave revolt led to the beginning of a new order. It rid the colony of slavery and brought with it political independence – the island was renamed Haiti. But the price was high – the only industry worth anything was sugar and this was destroyed; the white planter class was murdered and decades of civil war followed.

Progress had been rapid: by the time of the rebellion, St Domingue was producing nearly 80,000 tons of sugar per year and had an immense slave population, the highest of any colony – 480,000. There had been a number of attempted coups before 1791 yet the planters had failed to take heed of these warning signals. In 1750 a slave called Macandal, who had lost his arm in a sugar mill, became leader of the maroons. He attacked the white population by poisoning them. He was caught and burned alive in Limbé in 1758, but the campaign he had started continued without him.

On 20 August 1789, just over a month after the storming of the Bastille had marked the dawn of the French Revolution, the new French National Assembly stated that France had accepted the Declaration of the Rights of Man: 'all men are born and continue free and equal to their rights'. This statement was so strongly opposed by the planters and white officials in St Domingue that on 20 March 1790 the National Assembly passed a motion that said it was never the intention of the Assembly to put forward declarations that affected the government of the colonies. It was this blatant reversal of the Declaration of the Rights of Man that acted as a catalyst for the 1791 uprising. The original leaders were swiftly executed, an act that was widely condemned by the National Assembly back in France: on 15 May 1791 a motion, composed by Abbé Gregoire, was passed. It dealt with Creoles and stated, 'that the people of colour resident in the French colonies, born of free parents, were entitled to as of right and should be allowed the enjoyment of the privileges of French citizens, and among others those of being eligible to seats both in the parochial and colonial assemblies'. The planters were outraged. The situation came to a head on 23 August.

The slaves on the plantation of Count Noé, thirteen kilometres from the capital Cap Francois on the north coast, revolted. Simultaneously another three estates rebelled and within a few hours the northern part of the colony was under African control. The situation was complicated by the ambiguous status of the Creoles, who were half-white, half-black. They had more authority than the blacks, but were despised by both the white planters and the black slaves. They fought against the whites in the south. When they received no support from the black slaves, they reached an agreement with the plantation owners. As soon as news of the revolt reached France, Abbé Gregoire's motion was rescinded and, this time, the Creoles joined forces with the black slaves. Troops were sent over, but by this stage 2,000 whites had been killed and 180 factories and five plantations destroyed.

Two French commissioners were quickly dispatched to Jamaica to seek the help of the English. They were told that the government would provide 'every assistance and succour which was in its power to give'. An Englishman accompanying the commissioners to Cap Francois on 26 September 1791 wrote:

The first object which arrested our attention as we approached was a dreadful scene of devastation by fire. The noble plain adjoining the Cape was covered with ashes and the surrounding hills, as far as the eye could reach, everywhere presented to us ruins still smoking, and houses and plantations at the moment in flames. It was a sight more terrible than the mind of any man, unaccustomed to such a scene, can easily conceive – the inhabitants of the town being assembled on the beach, directed all their attention towards us, and we landed amidst a crowd of spectators who, with uplifted hands and streaming eyes, gave welcome to their deliverers (for such they considered us) and acclamations of *vivent les Anglois* resounded from every quarter.

The English kept their word and sent troops to St Domingue. However, their army was decimated by yellow fever: of their 15,000-strong force, 11,000 died.

A charismatic, middle-aged man, whom many women are said to have found attractive, Toussaint L'Ouverture had become leader of the Africans. He had been a coachman and storekeeper on a sugar plantation near Cap Haitien and could neither speak nor write French but was able to read it. He was well aware of the revolutionary principles of Liberté, Egalité, Fraternité.

L'Ouverture defeated what was left of the English army, ordered the ex-slaves back to work and agreed to protect the lives of the remaining whites. He claimed independence for the island on 7 July 1801 and awarded himself the title 'General for Life', calling himself the Bonaparte of France. Napoleon had already both granted and disallowed emancipation in the space of a few years; now he was determined to restore French power at any price. Neither must it be forgotten that his empress, Josephine, was the daughter of a planter in Martinique and had been brought up on her family's estate. He sent an army to the colony under the command of his brother-in-law, General le Clerc, whose orders were to adopt underhand means to defeat L'Ouverture, by first persuading the ex-slaves that slavery would never be reimposed and confirming L'Ouverture as leader, before gradually eroding his power. Le Clerc succeeded and managed to trick L'Ouverture, kidnap him and send him to France. He was thrown into the Fort de Joux prison in the Jura Mountains where he wrote semi-illiterate letters to Napoleon, who ignored him. Perhaps because he was instructed to, the jailer left for four days and L'Ouverture died of cold and hunger in his prison cell in 1803.

When he heard of this betrayal, L'Ouverture's deputy, Dessalines, exacted a horrific revenge, murdering every remaining white and Creole on the island. Dessalines promoted his general, Henry Christophe, to governor of the northern region of the island. Christophe was an extraordinary man. Born in 1767 on a plantation in either Grenada or St Kitts, at the age of twelve he ran away but was captured and put on a French ship bound for St Domingue. The naval officer who owned him sold him to a free black, the proprietor of a hotel in Cap Francois, where he became a stableboy and waiter. By the time he was twenty he had saved enough tips to buy his freedom. In 1794 he enlisted with L'Ouverture's army and within seven years had become a general.

By 1804 Dessalines was crowned emperor Jean Jacques I, but was murdered in 1806, and in 1807 Christophe declared himself president; by 1811 he'd appointed himself King Henri I of Haiti. At his coronation four princes, seven dukes, twenty-two counts, thirty barons and forty knights were present; he was crowned by a Roman Catholic priest.

To begin with Christophe was an incredibly successful leader. An impressive figure, six foot tall, well spoken in English and literate, he created a new coinage and a set of laws called Code Henry. He corresponded with fellow monarchs and tried to change the language on Haiti to English. He managed to persuade the people to return to work and restored the sugar industry to some extent. Gradually economic recovery began to take place. Christophe was in touch with William Wilberforce, the great emancipation leader, whom he asked to send him seven schoolmasters, a tutor for his son and seven professors specialising in classics, medicine, surgery and pharmacy. He had seven palaces built, one of which, Sans Souci, was said to be one of the finest buildings in the Americas: it had a library and a stream flowed beneath it that kept it cool.

The last castle he built was his undoing. He built La Ferrière on the top of a 914m peak, a few kilometres from Cap Haitien. The stones for the walls, which were thirty metres high and nine metres thick, had to be dragged up the mountain. It was said that La Ferrière could hold ten thousand men. But this final palace obsessed him and he lost his hold on reality, becoming irrationally cruel. To impress an English visitor he marched a company of soldiers across the parade ground and over the edge of a precipice – to show how well disciplined they were. In 1820 the army threatened to revolt. Half paralysed by a stroke and mentally imbalanced, Christophe shot himself with a gold bullet. Since then La Ferrière has been abandoned.

After his death, sugar production plummeted to its lowest level: 2,672 tons per year. Although keeping the sugar-cane industry in production was probably the last thing on most people's minds following the 1791 revolt, their newly found freedom and changes of rulers, at the time it was the island's sole means of raising revenue. Deerr comments:

> It is unfair to charge as a debt to the African race their failure to continue sugar production in what had been the most prosperous and wealthy of all the sugar islands . . . it is too much to have expected a race held in servitude, under conditions always severe and sometime ferocious, to have once bridged a gap which had only been crossed by Nordics after centuries of accumulated experience, and even then, as far as sugar was concerned, built upon a foundation of Arab experience.

Independence was finally recognised by Britain in 1825, and by France in 1838. But, in spite of Haiti's problems, it was a shining beacon of hope for slaves still in captivity.

That it was cruel and unfair to enslave a human being, depriving them of their freedom, was first recognised by the Quakers. George Fox, their founder, visited Barbados in 1671. Not yet in favour of the abolition of slavery, he wrote, 'I desired them . . . to deal mildly and gently with their negroes, and not to use cruelty towards them . . . and that after certain years of servitude they should make them free.'

Unsurprisingly, since so many religious orders had carried sugar cane throughout the world, many owned slaves. Of all the religious orders, the Jesuits had the most; on St Domingue alone they possessed 2,000 slaves. By the middle of the eighteenth century they had become the largest individual producers, shippers and slave-owners in the world. Deerr, however, believed that they were not inhumane and said, 'they worked to alleviate the lot of the slave and acted as a barrier against any excesses of the general slave-holding community'.

It may seem to be a conflict of interests for the great religions of the world to own slaves, but at the time many religious orders justified slavery on the grounds that it was a perfect opportunity to convert the heathen to Christianity. There was some argument over whether a person who had been baptised could continue to be a slave. William Fleetwood, Bishop of St Asaph and later Ely during the seventeenth century, said in a sermon, 'They [the planters] are neither prohibited by the Laws of God nor those of the Land from keeping Christian slaves. Their slaves are no more at liberty after they are baptised than before.'

He continued his defence by quoting St Paul, 'Let every man abide in the same calling wherein he was called.' Not everyone agreed: the French priest Père Lâbat once asked an English clergyman why the slaves were not baptised. He replied that it was considered wrong to keep a Christian soul in slavery.

The man who almost single-handedly stopped the slave trade and brought about the end of slavery in Britain and her colonies was deeply religious. William Wilberforce also saw in the slaves an opportunity for conversion, but this desire was tempered by a belief in the inalienable right of human beings to be free. His biographer, Oliver Warner, describes Wilberforce's life: 'Although he did not work alone, he was the mainspring of a fine movement, and it was his political skill, his personal gifts and charm, his enduring persistence, which in the end defeated the massed and vigorous forces of vested interest and obstruction.'

William Wilberforce was born in 1759 and spent almost his entire life fighting for the abolition of slavery. But, to begin with, no one would have predicted such a future for him. He grew up in Hull in a wealthy and frivolous family. Small, frail and with weak eyesight, he was uncannily serious for a child of a family that was both superficial and rich. His father died when he was nine and he was sent to live with an aunt and uncle in London. His mother eventually took him away from them, believing they were making him excessively pious, and she and the rest of his family attempted to liven him up. He wrote of this time:

> The theatre, balls, great suppers and card parties were the delight of the principal families of the town. This mode of life was at first distressing to me, but by degrees I acquired a relish for it, and became as thoughtless as the rest. I was everywhere invited and caressed. The religious impressions which I had gained at Wimbledon continued for a considerable time after my return to Hull, but my friends spared no pains to stifle them. I might almost say that no pious parent ever laboured more to impress a beloved child with sentiments of piety, than they did to give me a taste for the world and its diversions.

This lifestyle continued when he went to Cambridge University; he was encouraged to be louche and lazy, not only by his fellow students, but also by his own tutors, for they said that, since he had a personal fortune, he had no need of a degree or employment. But even at this age (he was only seventeen) he was moved to write a letter to the editor of the *Yorkshire Gazette* on the subject of slavery.

After university he won a parliamentary seat in Hull and kept a large house in Wimbledon. His life was a continuous round of parties and dinners; he loved to eat and dined on delicacies such as turtle, venison and asparagus.

And then something happened that was to fundamentally change Wilberforce and his way of life. The catalyst for the change was a man named John Newton, the rector of St Mary Woolnoth in London, who was introduced to Wilberforce by a Cambridge friend Isaac Milner. Newton was 60 and Wilberforce had turned 25, but the meeting was so momentous for Wilberforce that, although he was still admired for his wit, charm and social grace, he left his gentlemen's clubs, disassociated himself from his privileged set and acquired – or regained – a deep religious conviction. More than this, he closed his house in Wimbledon as being a needless expense and calculated that he could donate a quarter of his income to charity.

What was so surprising about the charismatic Newton was that he had been a slave-ship captain. Born in 1725, he was the son of a ship owner, who traded in the Mediterranean. Newton was press-ganged into service on a ship at the age of seventeen and endured five years of misery before he escaped. When he was recaptured, he was flogged and exchanged with a sailor on a slave ship. As a result of his troublesome behaviour, he was discharged on the condition that he work for the owner of a slave ship, a man named Clow. Clow and his black wife virtually enslaved Newton, making him labour alongside their slaves in a lime plantation without pay, clothes and with little food. Yet, during this time he managed to master six books of Euclid, practising drawing geometrical diagrams in the sand with a stick. He was released after a year when a visiting Englishman embarrassed Clow by commenting on his white slave. Newton was rescued and brought back to England by the captain of one of his father's ships who'd been told to look out for him.

Newton converted to Christianity around this time, and also started working on slave ships. For the captain of a slave ship he was quite a fair man, but he initially seemed to have no notion of the cruelty that was being inflicted on his captives. Gradually his opinion changed.

In 1788, he wrote *Thoughts upon the African Slave Trade*, in which he said:

> I should have quitted [the slave trade] sooner had I considered it as I now do to be unlawful and wrong. But I never had a scruple upon this head at the time; nor was such a thought ever suggested to me by any friend. What I did I did ignorantly; considering it as the line of life which Divine Providence had allotted me.

But in his old age he was haunted by memories of a 'business at which my heart now shudders'.

Newton not only had a profound influence upon Wilberforce, but on the country. In a committee of the House of Commons established in 1790 to investigate the slave trade, he said of the Africans, 'With equal advantages they would be equal to ourselves in point of capacity. I have met with many instances of real and decided natural capacity amongst them.' He added that assimilation with Europeans, far from 'civilizing these savages', had a poor effect on their morals: 'The most humane and moral people I ever met with in Africa were on the River Gaboon and at Cape Lopas; and they were the people who had the least intercourse with Europe.'

Newton was ordained at the age of 39 and appointed as curate of Olney. Shortly afterwards the poet William Cowper came to live in the town. He had been in an asylum for the mentally imbalanced and, as his madness returned, Newton took him in and cared for him for thirteen months. The two created a collection of songs often referred to as the Olney hymns. Altogether Newton wrote 280 hymns in his lifetime.

Wilberforce was helped in his endeavour by Granville Sharp. Sharp by name, he was also sharp in character and appearance, possessing a thin, pointed nose, prominent, almost skeletal cheekbones, a grim slit of a mouth and a chin that was practically dagger-shaped. In contrast, Wilberforce appeared reticent and shy, almost foppish. In a watercolour of him at the time painted by George Richmond he wears dark velvet pantaloons and shiny, patent leather shoes. He is contorted in his chair as if withdrawing from the onlooker, yet he cranes forwards with his head on one side, his eyes tiny, dark and bright, shining with an expression of interest and benevolence.

Sharp had been running a campaign to make slavery illegal in England. It had begun when a Barbados lawyer, Lisle, beat a slave severely and left him to die in the streets. Sharp found the man, took him home and nursed him back to health. Lisle then kidnapped the slave and shipped him from England to the West Indies, where he was sold. Sharp protested and Lisle charged him with stealing his property. There was a public outcry but, to Sharp's disappointment, Lisle dropped the court case. The next time a similar case came up, the court itself freed the slave, but on the grounds that the former master had attempted to seize him without a warrant, not, as Sharp would have liked, because the African was a free man once on English soil.

The third case, however, proved to be momentous. James Somerset, a fugitive slave, was brought before Lord Chief Justice Mansfield in 1772. Warner writes: 'It was clear that Mansfield, reluctant as he was to deal a blow to property, sympathised with long-established notions of freedom on English soil, and when, on 22 June, he delivered his judgement, he made history.' Mansfield said, 'Tracing the subject to natural principles the claim of slavery never can be supported. The power claimed never can be supported. The power claimed never was in use here or acknowledged by the Law,' and he concluded by stating that 'as soon as any slave sets foot in England he becomes free'. This ruling did not change the situation in England's colonies, but it boosted the Abolition movement and, above all, it affected the 14,000 slaves living in England at the time.

Wilberforce began to collect evidence against slavery and in 1787 he entered the parliamentary fray. He was invited by his firm friend William Pitt (the Younger), the prime minister, with whom he'd been at Cambridge, to put forward his argument to the House of Commons. It was going to be difficult: the public were not yet fully behind Abolition; there were considerable vested interests at stake; the king did not support it, neither did much of the Commons and certainly most of the House of Lords were against the idea.

And then, at the start of the campaign, Wilberforce fell ill.

By coincidence, Wilberforce had been at the same Cambridge College as Thomas Clarkson. While they were at university, Clarkson entered an essay-writing competition entitled 'Is it right to make men slaves against their wills?' He won the competition and, now a clergyman, he proved invaluable to Wilberforce, who was unable to muster enough energy, never mind evidence, to present his case. Clarkson's

approach was to visit places associated with the slave trade and, naturally, he picked Bristol to begin with. He wrote:

> On turning the corner within about a mile of that city at about
> eight in the evening, I came within sight of it . . . The bells of
> some of the churches were then ringing . . . the sound filled me,
> almost directly, with a melancholy for which I could not
> account. I began to tremble, for the first time at the arduous task
> I had undertaken, of attempting to subvert one of the branches
> of the commerce of the great place which was before me.

In spite of his fear, he hung around the taverns and docks, speaking to sailors and captains, risking the wrath of merchants, slave-ship
owners and plantation owners who called him a 'white Jacobin nigger'.
Pinney, whose house was but metres away from the docks, said that
alongside the borer worm, abolitionists were the most dangerous pests
threatening his plantations. Clarkson collected shackles, thumb-
screws, mouth-openers and other instruments of torture used on
board ships and recorded eyewitness testimonials.

In 1788 Wilberforce collapsed with severe intestinal trouble.
Neither his doctors nor his friends believed he had the strength to
recover. He went to Bath in an attempt to restore himself at the town's
spa and wrote to Pitt asking him if he could put a resolution forward
to the Commons to regulate the slave trade. Pitt was in an awkward
position; proposing a motion even to curtail, never mind abolish, the
slave trade would put his career in jeopardy. Nevertheless, wrote
Wilberforce, 'With a warmth of principle and friendship that has made
me love him better than ever I did before, Pitt gave his promise.'

Clarkson's research proved invaluable. Liverpool merchants were
forced to admit that 10 to 15 per cent of their slaves died during the
Middle Passage – a grave underestimate, as Wilberforce was to show
later, but sickeningly large enough to move Pitt to considerable anger.
He said in Parliament:

> If the Trade cannot be carried on in a manner different to that
> stated by the honourable members opposite to me, I will retract
> what I said on a former day . . . I will give my vote for the utter
> abolition of a Trade which is shocking to humanity, is abominable
> to be carried out by any country, and reflects the greatest

dishonour on the British senate and the British nation. The Trade, as the petitioners propose to carry it on, without any regulation, is contrary to every human, to every Christian principle, to every sentiment that ought to inspire the breast of man.

Lord Edward Thurlow, loyal to the king and anti-Abolition, called the bill 'a five days' fit of philanthropy' and stated that merchants would be ruined. Most of the House of Lords agreed with him. Pitt, now seriously angered at the lack of support from his own party, told Sharp that if he were defeated in the Lords he could not remain in the same cabinet as his opposers. Fortunately, he won. By two votes.

But this was nothing like the victory that Wilberforce and Pitt had envisioned. More evidence was required before the Lords were willing to determine *how* the slave trade was to be regulated.

Wilberforce, helped by a hearty dose of opium and a trip to the Lake District, returned invigorated and continued to campaign. In the colonies the slaves sang, 'Mr Wilberforce for Negro, Mr Fox for Negro, the Parliament for Negro, God Almighty for Negro.' In 1789 the impassioned Wilberforce gave one of his most eloquent and powerful speeches in support of free men working in the West Indies in England's sugar-cane plantations.

Sir, the nature and all the circumstances of this Trade are now laid open to us. We can no longer plead ignorance. We cannot evade it. We may spurn it. We may kick it out of the way. But we cannot turn aside so as to avoid seeing it. For it is brought now so directly before our eyes that this House must decide and must justify to all the world and to its own conscience, the rectitude of all the grounds of its decision. A society has been established for the abolition of this Trade, in which Dissenters, Quakers, Churchmen – in which the most conscientious men of all persuasions – have united and made common cause. Let not Parliament be the only body that is insensible to the principles of natural justice. Let us make reparation to Africa, so far as we can, by establishing a trade upon true commercial principles, and we shall soon find the rectitude of our conduct rewarded by the benefits of a regular and growing commerce.

What helped him at the time were two things: first, the vicissitudes inflicted on *white* men in connection with the slave trade. As

Wilberforce pointed out, 'More sailors die in one year in the Slave Trade than die in two years in all our other trades put together.' Secondly, aided by the Anti-Saccharite Society, the tide of public opinion started to turn against the slave trade. This organisation was founded by William Fox, a Christian who had also set up the Sunday School Society, and who advocated buying sugar from sources other than the English colonies in the Caribbean.

> To abstain from the Use of Sugar and Rum until our West Indian planters themselves have prohibited the importation of additional slaves, and commenced as speedy and effectual subversion of slavery in their islands, as the circumstances and situation of the slaves will admit; or until we can obtain the produce of the sugar cane in some other mode unconnected with slavery and unpolluted with blood ... A family that uses 5lb [11.34 kilos] of sugar per week with a proportion of rum will, by abstaining from the consumption for 21 months, prevent the slavery or murder of one fellow creature.

Fox went on to describe how an agent had found the whole body of a 'toasted negro' in a cask of rum to help flavour the beverage. The East Indian Company was quick to take advantage of the Anti-Saccharite Society's publicity and produced sugar bowls with the slogan, 'East India Sugar not made by slaves'. (The terrible irony is that, while Indian sugar was not made by African slaves, it was made by *Indian* slaves, particularly sugar from Bengal.)

Wilberforce continued to amass evidence and set it before the House of Commons. The next date for a debate was April 1792. Wilberforce related how six slave ships had anchored off the African town of Calabar. The African dealers had asked for a higher price than usual. In response, the captains of the ships opened fire for three hours, injuring many and killing twenty. No one returned fire. At the end of this massacre, the dealers, unsurprisingly, agreed to sell at any price.

Nevertheless, MPs still argued for a gradual end to the slave trade. Fox retorted: 'I believe the Trade to be impolitic, I know it to be inhuman. I am certain it is unjust. I find it so inhuman and unjust that, if the colonies cannot be cultivated without it, they ought not to be cultivated at all.'

The debate continued throughout the night. Pitt spoke just before dawn. He summed up his speech by saying:

What astonishing, I had almost said, what irreparable mischief have we brought upon that Continent? How shall we hope to obtain, if it be possible, forgiveness from Heaven for the enormous ills we have committed, if we refuse to make use of those means which the mercy of Providence has still preserved us for wiping away the shame and guilt with which we are now covered?

At 7 a.m. a vote was finally taken. The decision of the House was that the slave trade should be gradually abolished, ending four years later in 1796. It was not entirely unexpected and Wilberforce, a politician since his early 20s had always known that reform would be slow. But the 1796 date was to come and go and it would be many years before the trade was even regulated, let alone eradicated. Both King George III and the Lords contested the bill; more damaging still, in 1793 war with France broke out. Sir Reginald Coupland, author of a biography on Wilberforce, wrote: 'Year after year now, it was less and less possible for English ears to hear the far, faint cry of Africa beyond the guns in Europe.'

Thomas Dowling, the surgeon on board the slave ship *Recovery*, had told Wilberforce about the captain John Kimber. After Wilberforce in turn mentioned Kimber's name and recounted during the April 1792 debate the incident he'd been told, he was threated by Kimber. A fourteen- or fifteen-year-old slave girl on the ship had a venereal disease and could not eat; Kimber had her suspended by her hands and then by each leg while he flogged her to death with a whip. Kimber was arrested six days later and tried, but the jury acquitted him of murder and prosecuted Dowling and a second witness, Stephen Deveraux, the third mate, for perjury. One of Wilberforce's friends, Lord Rokeby, was so concerned about the nature of Kimber's threats that he insisted on accompanying Wilberforce to Yorkshire with a pistol in his pocket.

After eighteen years in power, Pitt left the government, and Henry Addington, who was a Conservative and not in favour of Abolition, became Prime Minister. Wilberforce was beaten three times more in Parliament, and admitted that the likelihood of the slave trade ending seemed weaker than when he'd first begun to fight for Abolition in 1787. He wrote, 'Hope deferred maketh the heart sick.'

Pitt returned to power in 1804, but died in office in January 1806. Ironically, his death helped the cause, for the new ministry included

Fox and Granville Sharpe. Finally, on 10 June 1806, a resolution was passed both in the Commons and the Lords to prohibit the sale of slaves to British competitors. On 23 February 1807 a bill for the complete abolition of the slave trade was passed. Wilberforce was given the biggest standing ovation awarded to any man in Parliament. As he sat with tears streaming down his face, he said it was his most triumphant hour. He had been waiting for this moment for twenty years.

It was a tremendous victory and long overdue, but did not change the situation for the thousands of slaves already in the colonies. Arguably the situation became worse as plantation owners increasingly took out their frustration on their captives. In Demerara, in British Guiana, the slaves heard of the 1807 act and believed that they had been freed. Some refused to work; the matter was resolved when troops were called out and five were lashed a thousand times. The news reached London in a distorted form: the slaves had risen up en masse and murdered the planters. Evidence of further atrocities continued to accumulate, spurred on by testimonials such as Olaudah Equiano's. Wilberforce was roundly abused by the popular press. He responded, 'They charge me with fanaticism. If to be feelingly alive to the sufferings of my fellows is to be a fanatic, I am one of the most incurable fanatics ever permitted to be at large.'

Wilberforce had believed that, once the slave trade ended, emancipation would gradually happen without any need for legislation, but it became clear that he had been wrong. He was thwarted by George Canning, who was foreign secretary and would become prime minister in 1827. Canning did not support emancipation and said, 'I abjure the principal of perpetual slavery; but I am not prepared now to state in what way I would set about its abolition.' Instead he did his best to instigate and pass laws that ameliorated the lot of the slaves. For example, he said that they should be allowed to own property, they could marry and women were not to be whipped. None of this changed the fact that the slaves remained in bondage, subject to the only slightly reduced whim of their masters. Yet the result was furious opposition led by the Assembly of Dominica, which called on all colonies:

> to combine our Efforts and to Energetically mark our firm determination never to consent to kiss the rod, or meekly lick the hand just raised to shed our blood; but with one Voice to denounce in the Face of the World the blind fanaticism of the

'Saints' [Evangelicals who opposed slavery and included Wilberforce] who would now for a Phantom cast to Perdition these once valuable and highly valued Colonies, while at the same time they are looking on with cold bloody apathy to the miseries of Ireland and their own Poor.

On 1 June 1824 Wilberforce appeared in the Commons once more, filled with indignation over the case of John Smith, a missionary in Demerara, where the slaves, as elsewhere, believed that freedom had been granted to them. A number, led by Jacky Reed, plotted an uprising, and the insurrection broke out on 18 August 1823, with over 9,000 slaves from 37 estates involved. The uprising was easily suppressed and 47 slaves were executed.

The ringleaders had all been members of a Congregationalist chapel whose pastor was Smith and what incensed Wilberforce was that Smith was accused of promoting and exciting them to rebel. He was sentenced to death without trial, but, before he could be killed, he contracted tuberculosis in prison and died. His widow was forbidden to attend his funeral.

Canning believed him guilty and said, 'I have no difficulty in stating the honest conviction of my own mind to be this, that of that crime . . . which consists in the silence of Mr Smith upon . . . a danger which he knew to be imminent. I cannot acquit Mr Smith.' Evidence may have been falsified to obtain this sentence, but it seems clear that he was not guilty. Jacky Reed had written to Smith intimating revolt; this was Smith's reply:

To Jacky Reed,

I am ignorant of the affair you allude to, and your note is too late to make any inquiry. I learnt yesterday that some scheme was in agitation; without asking any questions on the subject I begged them to be quiet. I trust they will; hasty, violent and concerted measures are quite contrary to the religion we profess, and I hope you will have nothing to do with them.

Yours for Christ's sake,

J.S.

Wilberforce used the incident to debate the abolition of slavery again on 15 June. By this stage abolitionists were in the majority in the general population; between 1826 and 1832 more than 3,500 anti-slavery petitions were submitted to the House of Lords alone.

However, it is more likely that economic arguments held sway with the government. Adam Smith, in *An Enquiry into the Nature and Causes of the Wealth of Nations*, wrote: 'From the experience of all ages and nations I believe that the work done by slaves is in the end the dearest of any.' By 1830 Britain itself had also sunk into an economic crisis. Factories closed down, unemployment rose, wages declined; at least 10 per cent of the population were living in poverty. Historians, including Dr Eric Williams, author of *Capitalism and Slavery*, who was Trinidad and Tobago's Prime Minister between 1959 and 1961, believe this was an inevitable consequence of a country transforming itself from an agricultural system of production to an industrial one. By this stage more than half the population lived in urban areas, whereas only sixteen years before the majority had lived off the land. The abolition of slavery seen in this context was less a humanitarian gesture and more a part of Britain's integration into a modern capitalist economy. Britain was on the brink of the industrial revolution, metamorphosing into a nation that would build its economy on the strength of its trade and manufacturing capability. Arguably the colonies and their out-dated methods of producing sugar had become less important to this new order.

As for Wilberforce, his energy and his life were almost spent. He has been described as a persuasive and eloquent speaker: 'the nightingale of the House'. Williams depicted him more harshly:

> As a leader he was inept, addicted to moderation, compromise and delay. He deprecated extreme measures and popular agitation. He relied for success upon aristocratic patronage, parliamentary diplomacy and private influence with men in office.

True as this may be, it is unlikely that slavery could have ended earlier given the amount of men in government and among the aristocracy who directly benefited from slavery and the slave trade. It was only as the political and economic climate changed that Abolition was finally achieved. The case for emancipation was carried, although others led it, for Wilberforce fell critically ill. On his deathbed he heard that the measure had been passed, and, a year after he passed away, the Act was enforced. Coupland wrote: 'At midnight on 31 July 1834, eight hundred thousand slaves became free. It was more than a great event in African or British history. It was one of the greatest events in the history of the world.' From the beginning of colonial history until that date, twenty million Africans had been enslaved; two-thirds of them died because of our desire for sugar.

Abolition did not, however, lead to a swift change of fortune for a people who had been dislocated from their country, culture and compatriots, nor for those who had been born into and had never known anything but slavery. To begin with, only children younger than six were freed. Older slaves had to work through a six-year apprenticeship, though this was abandoned in 1838. The plantation owners themselves were compensated for the loss of their slaves, receiving £26 per person – leaving an excellent historical record regarding the number of slaves, as well as who owned them. As Deerr said:

> An examination of the list of slave owners shows how great was the representation of the armigerous, titled and landed classes on the one hand and of the great merchant houses of London and Liverpool on the other, demonstrating the great influence of the West Indian interest in both Lords and Commons. At the other end of the scale are many instances of the ownership of but one slave, indicating that the odium of slave ownership ran through all classes of the community and that it was not confined to any one social order.

On islands such as Barbados that are small and were wholly devoted to sugar-cane plantations, there was nowhere for the ex-slaves to go. The planters brought in the Tenantry System to enable them to push wages down by a further 25 per cent: under this regime the workers were allowed to remain in their huts and use a plot of land but had to work exclusively for the planter. They now had to work harder for reduced wages. Plots were usually only a tenth of a hectare and workers could not afford to buy more land, even if there had been some spare. During the nineteenth century sugar prices decreased and planters further reduced rates of pay. As Sidney Mintz, author of *Sweetness and Power*, points out, the planters sought to recreate pre-emancipation conditions using hunger instead of the discipline of slavery.

Where ex-slaves could leave, on larger islands like Jamaica, they did. It was not merely a question of money; a person who had been sold into slavery to grow sugar cane was unlikely to want to continue this profession when free. Here the ex-slaves eked out a living from small scraps of land, managing to subsist, rather than work for their former masters.

This, of course, created a problem for the planters, who lost their workforces. Britain had founded a colony on Sierra Leone for slaves the Royal Navy had captured at sea as they attempted to suppress the slave trade. Plantation owners suggested that, in return for partially paying their fare, these people could work as indentured servants and would be given wages after a fixed period of time. Naturally, few took them up on this offer; the British public and the Africans felt it sounded too similar to slavery. Instead the plantation owners put the same offer to the governments of China, Japan and India, who agreed to the terms. Sadly, this turned out to be slavery under a different name. It was almost impossible for these new nineteenth-century slaves to receive wages when their period of indenture was over, or a free passage back as they'd been promised.

Emancipation did not affect all colonies at the same time. Britain freed slaves in 1834; France and Denmark emancipated their slaves in 1848; the Netherlands in 1863, Puerto Rico in 1873 and Cuba in 1884. But America had to fight a bloody civil war before Abolition was achieved. This situation, where some colonies had slaves while others did not, together with the shift to modern capitalism, resulted in some difficult issues, especially for the British Government.

Sugar consumption had risen to such an extent and become so important in terms of the British Government's income (because of the duties paid on it) that it could no longer allow just the colonial plantation owners to provide Britain's sugar. What the government wanted was the cheapest sugar to reach the widest markets. This might mean buying sugar from colonies that still kept slaves, rather than from British colonies – free trade, in other words.

Free trade did indeed force down the price of sugar and, as a result, increase the markets and raise sugar consumption. By 1800 British consumption had risen by 2,500 per cent in 150 years; 248,932 tons of sugar were sold worldwide. By 1830, before beet sugar had reached the world market, total world production was 581,179 tons, and by 1890 world production exceeded six million tons. Free trade also finally changed the plantation system, although, as Deerr pointed out, this was almost a century after emancipation.

Though the West Indian Company gave planters money to establish central factories on each island, the impact of sugar cane on these islands was dramatic. Most of the population – almost 95 per cent of Jamaica, Haiti and those people on the coast of Brazil – are descended from slaves, who were themselves a racial mixture from diverse parts of Africa giving the islands a high degree of ethnic diversity. But, as Jock Galloway points out, other structures persist. There is rigid racial and social stratification connected with occupational status; racial groups are segregated, and most of the population are lower class and poor. For instance, in Barbados, the undesirable social conditions that still prevail are a heritage of colonial times.

Ex-colonies produce less sugar than areas with the same climate and environment but that have not had slavery, leaving him to conclude that the 'colonial plantation system has resulted in a social structure which is also prejudicial to efficient sugar production'.

The heyday of sugar production in Barbados was in 1966, the year it gained its independence, when the country produced just over 200,000 tons of sugar a year. Now it barely manages 30,000 tons. The fact that sugar dropped dramatically when there were no white foreigners running the plantations indicates that there was still a seething undercurrent of resentment, a legacy from a crueller era. But the reason why sugar production has sunk so low recently has less to do with slavery and a lot more to do with Andreas Maggraf's sugar-beet discovery in the mid-eighteenth century, the development of which would enable sugar to be produced on European soil instead of in Europe's former colonies.

6. THE FLOWERING OF THE WHITE TRANSPARENT

> We are at least freed from the vain search for the undiscovered
> and undiscoverable essence of the term species.
>> Charles Darwin, *On the Origin of Species*, 1859

For sugar-cane growers the nineteenth century was a struggle. Abolition
was, of course, the greatest change to sweep through the plantation
system, and the introduction of free trade contributed to their financial
difficulties. But there were other problems. One of these was from sugar
cane itself; another was from the rise in sugar-beet production.

The Creole cane had been planted throughout the world, only to be
largely replaced by the Otaheite cane during the late eighteenth
century. Also around this time, sugar-cane growers witnessed the discovery of two new types of cane. The first was Cheribon, also know as
Transparent, Preanger, Crystalina and Batavian. It occurred in several
mutations: purple, yellow and striped. The most successful of these
was called White Transparent. The Cheribon was taken to Savannah,
Georgia and then on to Louisiana at the beginning of the nineteenth
century before displacing the Otaheite in Cuba in 1840.

The second new variety was noticed back in 1796, but evidence
for its existence was not published until much later, in 1832. It was
found by James Duncan, a surgeon working for the East India
Company stationed at Canton. He sent some canes to William
Roxburgh, who was also employed by the East India Company and
who was collecting material for his *Flora of India*. Roxburgh agreed
with Duncan that this particular cane, which was distributed
throughout northern India, was different from other varieties and
named it *Saccharum sinense* – the Chinese cane. No mention of it, by
the British at least, was made until 1848, when Arthur Crook gave
evidence before the Select Committee on Sugar and Coffee Planting.
He said there were three kinds of cane native to India and, according
to his experiments, the Chinese cane was the best. He added that he
had taken his original stock from the East India Company's garden.

But in mainstream cultivation there were now only three varieties, the Cheribon, the Creole and the Otaheite. These three were almost the same as each other and the older variety, the Creole, had remained practically genetically identical to the first canes that were spread by the Polynesians thousands of years ago. In effect, sugar-cane growers had inadvertently created a giant global monoculture. The Otaheite, however, whose replacement of the Creole had been so swift and dramatic, now displayed its fatal flaw – it was extremely vulnerably to red rot, the disease apparently foreseen by Buddha. In 1840, the Otaheite in Mauritius was all but wiped out and, over the next few years, the Otaheite's decline was as rapid as its rise: red rot devastated sugar-cane production throughout the Americas, finally reaching the West Indies in the 1890s.

Sugar cane is grown vegetatively – in other words, part of the stem is cut off and placed in the soil, where it sprouts new roots and buds. Therefore the new shoot will be identical to its parent. It was not until the late nineteenth century that anyone realised that sugar could be propagated from seed: no one had ever seen a sugar-cane flower, or, at least, had never mentioned their findings. The reason is partly because of this vegetative cultivation, and partly because the Creole produced sterile males. The plantation manger's bible, *The Practical Sugar Planter*, stated: 'No variety of sugar cane is known to perfect its seed (or to produce anything like seed).'

It was an employee of the Highlands estate in Barbados, Iran Aeus Harper, who recognised that sugar cane could be bred. In May 1858 Harper noticed some seedlings in one of the cane fields and told the owner of the estate. According to an article published in the *Journal of the Barbados Museum and Historical Society*, when asked how he could tell that it was sugar cane and not Guinea grass, which looks similar, Harper said, 'Run yuh finger 'long de edge of de leaf; de guinea-grass leaf doan have no sharp edge; de cane leaf do and dis one dat way.' The plantation owner, James Parris, wrote to the *Barbados Liberal*; his letter was published on 8 February 1859.

Dear Sir,

In accordance with your request, I now send you the following particulars regarding the canes established from the seed, and which are now growing at Highlands plantation.

I think it was somewhere in the month of May last year when my attention was called to the fact of there being several cane plants growing in a field of ratoons, which the superintendent pronounced as having been grown from the seed of the cane arrow. On the first examination I thought it was a mistake, they bore so close a resemblance to Guinea grass when it grows from seed; but, as there was not any kind of this grass growing on or near the field in question, I could not account for its presence there, and this circumstance called for a strict examination on my part, the superintendent declaring positively all the while that they were verifiable canes. After having satisfied myself that they really were canes, I caused all that could be found to be removed and transplanted to another field, but, in consequence of the weather being very dry, I could only save seven of them, and these are now alive and growing. I intend having the plants from these put in a spot by themselves this year, hoping to obtain seed from them again . . . It appears as if they are the seed from these kinds: the Bourbon [Otaheite], Transparent [Cheribon] and Native [Creole]; that is, the plants which are growing have the appearance of these at present.

And remain, Dear Sir

Yours respectfully

James W. Parris

His letter was copied by the *Markets Review* and the Australian press, but neither planters nor botanists paid any attention. The seedlings turned out to be from one of the Cheribon varieties, White Transparent. A planter named Drumm even sent some seeds to Kew via his British agent, Alfred Fryer of the Manchester Sugar Refinery, in 1871. The reply from Kew was terse, quoting Napoleon I's botanist, F. Tussac, 'It is often found in cane that there are many seeds but that they are sterile.'

One reason why this discovery was ignored was because it came from the wrong place: the descendant of a slave and a Barbadian planter had discovered and disseminated the knowledge, not an eminent botanist from the mother country. Secondly, science at the time was concerned with classification and categorisation – reproduction and heredity were the province of animal breeders, not scientists.

Then, however, Darwin published *On the Origin of Species*, his revolutionary new theory of natural selection, which helped change entrenched attitudes. In addition, the planters themselves became desperate for a breakthrough: their crops were devastated by disease and the only way they could save the industry was to breed new varieties that were disease-resistant.

In 1888 another man made the observation that sugar cane could indeed flower, and this time he was listened to, a fact that revolutionised the sugar-cane industry. In spite of his achievement, he is almost unheard of today and was little acknowledged in his lifetime. We have Jock Galloway to thank for unearthing the story of John Redman Bovell.

Bovell was born in 1855 in Barbados, the eldest of fourteen children. He was not from a particularly wealthy or renowned family; the Bovells had been in Barbados since the seventeenth century practising medicine, but had become linked to the planter class through marriage. Bovell had mediocre schooling: he attended school and the best college on the island, but did not win a scholarship and did not have enough money to go to university. Instead he became a sugar-estate manager until he suddenly switched jobs, becoming a superintendent of the Boys' Reformatory and Industry School at Dodds. Galloway believes he changed careers because his new position would allow him to pursue his great love: experimental research into sugar cane. The job at Dodds enabled him to farm 36 hectares of land and use the boys as a labour force.

Within two years of taking up this post, in 1885, the government awarded Bovell the position of superintendent of an experimental station at Dodds. It was merely recognition of work he was doing already and he continued to work for the boys' school, but the formal title helped him cultivate connections with Kew Gardens and other scientific organisations, such as the Linnean Society.

An undated picture of him shows him at an indeterminable age: he is balding with short white hair round the sides of his head, yet his appearance is youthful. With dark eyes, sculpted cheekbones, an aristocratic nose and perfect Cupid's bow, he could have been a movie star. It is this image of him that is currently on the Barbadian two-dollar note.

Galloway says that, in spite of his lack of formal training, he was a brilliant experimental scientist and, in 1888, Bovell and a colleague, John Harrison, published their 'rediscovery' of sugar-cane fertility. The *Argosy*, a Demeraran newspaper, wrote, 'A door, revealing a limitless vision, is opened to breeders and experimenters.' Now that the planters could read about his accomplishment in the newspapers, they praised Harrison and Bovell at the House of Assembly.

Barely a few years later, however, the planters tried to persuade Bovell to resign and attempted to undermine him; they presumably did not take kindly to being told how to grow sugar cane by an unqualified Barbadian. In the main sugar publication in America one planter wrote of their 'great dissatisfaction':

It appears that the government has allowed the Superintendent of the Station, who is only a sugar planter, and holds no diploma in science, to usurp the place of the Professor of Agricultural Science . . . What value, I venture to ask, can be attached to the research conducted by an unqualified man?

Bovell wrote frequently to Kew to complain about the planters' attitudes. In the meantime, he continued his research, which largely consisted of an exploration of the genetic variability within the Otaheite cane.

Now that he knew these could flower he allowed them to pollinate and set seed and then tested the resulting offspring. At first he let them pollinate by chance; this meant that he knew the female parent, but not the male. Then in the early part of the century, he would only allow known males to pollinate with known females by placing the canes inside a lantern. This is a tall structure with paper walls that could be wheeled into place and folded around a flowering cane and a male stem to prevent wind-blown pollen from other males contaminating the results.

Although growing canes vegetatively creates offspring that are almost identical to the parent, breeding between or within varieties, such as an Otaheite cane with a Creole, introduces greater variation. By the time he was 69 in 1924 Bovell had raised 118,669 seedlings. Very few would be commercially successful, but there was an exception: a variety he created in 1910 called BH [Barbados Hybrid] 10/12. It grew vigorously, had a high sugar content and was resistant to root diseases. Though it was susceptible to one disease – mosaic disease, which is caused by a virus and spread by aphids – by 1920 the new kind of cane was considered to be one of the best in the world.

After he retired, Bovell began to breed the wild strain of sugar cane with these noble canes. It was this line of research that was ultimately to lead to modern-day breakthroughs in the creation of new breeds of sugar cane. At the time, his experiments proved that, compared with his new hybrids and White Transparent, the Creole cane had the lowest yields of sugar by a long way and that in every year but one it had been attacked by fungal diseases. He showed that, by replacing White Transparent with his varieties on Barbados, revenue would increase by almost $1.5 million.

Kew finally responded and, as a result, Bovell was given a pay rise and a little more of the approbation he deserved. It thus took a British institution's approval and the almost total collapse of their sugar-cane production before a Barbadian discovery – cane breeding – was accepted by Barbadians. The year before Bovell retired his hybrids earned the planters an extra $2 million, although they were planted on less than two-thirds of the available land. By the 1930s sugar production had risen by 76 per cent. Bovell was equally successful elsewhere: the introduction of BH 10/12 to Puerto Rico increased the country's revenue by $10–12 million a year. As well as revitalising the sugar-cane industry through the biggest transformation of sugar cane since it was first domesticated, Bovell's real achievement was to lay the foundations for scientific research that was to culminate in the unravelling of the sugar-cane genome in the twenty-first century.

Bovell died in 1928. His work did not receive sufficient credit in his lifetime and after his death he slipped into obscurity. His colleagues, such as John Harrison, were knighted; there is not even a commemorative plaque at the school where he carried out his greatest work. The only physical memorial to him is his tomb, a few hundred metres away between the lilies and the frangipani trees in St Philip's churchyard. Yet, when I tried to find his grave among the headstones whose names were worn to dust, none of the Barbadians worshipping in the church or working in its grounds knew where it was, or even who the man on their two-dollar bill was.

The second great threat to sugar cane was the rise of beet sugar, a plant that could be grown in temperate lands and close to all major markets.

It was Andreas Maggraf's entrepreneurial student, Franz Carl Achard, who developed his tutor's innovation. He impressed Frederick William III of Prussia enough to be given land for a factory, which he opened in Silesia in 1801, and a ten-year monopoly on sugar-beet manufacture. There is a story that the English were so concerned about this turn of events that they attempted to bribe Achard to deny that sugar to rival cane sugar could be made from sugar beet. As Napoleon was the source of this rumour, it is difficult to determine whether it is true. Other sources suggest that the English bribed Continental refiners to put pressure on Achard. German police investigated the allegation at the time and Achard himself said to a former pupil that he had been offered 50,000 thalers in 1800 and 20,000 in 1802 by Continental refiners. In any case, Achard's factories failed to begin with: they lacked technical knowledge and practical experience. In the end, the manufacturers turned to grapes and were partially successful at producing grape syrup.

What changed the situation was the Napoleonic War: English command of the seas meant that the French were cut off from their West Indian sugar supplies and, as prices mounted, they turned back to beet. On 18 March 1811 Napoleon ordered the Minister of the Interior to take all steps necessary to encourage the growing of sugar beet and the erection of factories. Farmers were to be advised 'that the growing of beet root improves the soil, and that the residue of the fabrication furnishes an excellent food for cattle'. A week later, he took steps himself: at the Palace of the Tuileries he signed a decree that 31,970.5 hectares of land be planted with beets as quickly as possible and that six experimental stations should be established for the instruction of farmers and landowners. One million francs were poured into the project.

A man called Benjamin Delessert erected a factory at Passy and by 1812 he'd made a small quantity of well-crystallised beet sugar. Napoleon, on hearing this, was enraptured. According to a contemporary French writer, he said, 'We must see this. Let us go at once.' After viewing the results for himself – and presumably tasting them – he took off his Cross of Honour and pinned it to Delessert's chest. The next day he announced that 'a great revolution in French commerce' had been achieved.

By 1818 France had an annual output of four million kilos of sugar from beet. However, when the blockade against maritime trade was lifted at the end of the Napoleonic War, plantation owners from the colonies hastened to dump their own stagnating cargoes of sugar on Europe. The newly established beet factories, still struggling with primitive processing methods and low-quality raw material, were unable to withstand the onslaught.

Sugar beet was not finished, however. During the struggle for abolition of slavery, farmers in Europe had realised that it would be possible to promote their sugar as being slave free. In addition, sugar beet was a useful crop in other ways: the leaves and roots could be used for cattle feed; the incredibly long roots, over a metre in length, brought nutrients to the surface and improved the soil structure. Including beet in a crop-rotation scheme increased the yield of other crops.

Just as Bovell was managing to revolutionise sugar-cane crops, by the end of the nineteenth century, refining of sugar beet had improved to such an extent that their sugar was indistinguishable from refined cane sugar. Beet sugar was flooding into European markets; for example, two-thirds of the sugar sold to Britain in this period was from sugar beet. This obviously had a detrimental effect on every aspect of the sugar-cane industry, from cane cutting through to refining.

7. OUT OF THE STRONG CAME FORTH SWEETNESS

Marmalade of Eggs the Jews Way
Take the yolks of twenty-four eggs, beat them for an hour; clarify one pound of the best moist sugar, four spoonfuls of orange-flower water, one ounce of blanched and pounded almonds; stir all together over a very slow charcoal fire, keeping stirring it all the while one way till it comes to a consistence; then put it into coffee-cups, and throw a little beaten cinnamon on the top of the cups.

This marmalade, mixed with pounded almonds, with orange-peel, and citron, are made in cakes of all shapes, such as birds, fish and fruit.

Hannah Glasse, *The Art of Cookery Made Plain and Easy*, 1747

It is no exaggeration to say that two men changed the face of the sugar industry. Born within a year of one another, they would create a company so strong that it became a multinational whose name was known the world over and which grew to dominate global sugar trade for the best part of a century. Yet the two men who founded Tate & Lyle never even met.

Henry Tate was born in 1819 in Chorley, Lancashire, the seventh son of a Liverpool Unitarian clergyman, William Tate. At thirteen Henry was apprenticed to his older brother Caleb as a grocer. By the age of twenty he'd set up his own business and by 36 he had a chain of six shops. In 1859 he went into partnership with John Wright, a sugar refiner in Liverpool. Wright was himself a partner in Walker, Wright and Co, which was also a wholesale grocers as well as a small sugar business. He would also have come across other sugar refiners through his grocer shops as, at the time, grocers sold sugar loose. However, Henry was relatively shy and had no intention of taking a place in the civic community of Liverpool, neither was he inclined towards public life.

So what made Henry Tate, then a moderately wealthy businessman, reinvent himself as a sugar refiner? Henry's wife was thought to be his chief adviser and she may have played a role in his move from grocer to sugar baron. Born Jane Wigwall, she married Henry in 1841 and by 1859 they had eight children, although two died at a early age. Henry's obituary in the *Daily News* on 6 December 1899 credits his wife with a great deal: 'There are probably few things in the way of business that a shrewd Lancashire man with a pawky Scotch body for a wife cannot do . . . After the grocer's shops, Mrs Tate prompted her husband to the higher flights of sugar refining.'

Henry gave up his grocery business in 1861, two years after he'd become a sugar refiner, selling his shops to George Bird, second husband of his widowed sister-in-law, Hanna Tate (Caleb had died in 1846) – so the firm still remained with the family. The first ten years in refining were tough for Henry: on 18 September 1863 he wrote in a letter, 'our business is so miserable'. The refinery, of which little is known, probably produced bastards – low-quality sugar and treacle. Later he was to write, 'Our market is very flat and prices for our product very low. I really don't know what's to become of the trade.' Yet in 1869 he managed to buy out his partner John Wright and bring his sons into partnership (Alfred, then 25 years old, and Edwin, who was 22). Henry had had limited schooling himself (his father had taught him at home), but he made sure that his sons received a good education and training at work. Moreover, he had the facility for surrounding himself with highly competent men and insisted that even the humblest of his employees should be highly trained.

It was only after Henry had dissolved his association with John Wright that his business became profitable. The construction of a new factory at Love Lane in Liverpool was the second major decision of his life, after leaving the grocery business, and why Henry Tate proved to be such a successful sugar refiner can be seen in the way in which he ran that first solo refinery. According to the historian Professor Philippe Chalmin, whose extraordinarily detailed masters degree deals with the rise of Tate & Lyle, Henry was the first person to understand that 'the "sugar battle" could only be fought – and won – on the British side with a large-scale production of high-quality sugar'. This meant using the techniques perfected at the Walker's refinery in Greenock, Scotland, so Henry sent his son Edwin to Greenock; he reportedly came back 'dazzled'.

While Love Lane was being constructed, Henry Tate, in an example of what was to become his constant search for new technology, acquired the patent rights to a French process called Boivin-Loiseau (after the names of the inventors), which used lime and carbon dioxide gas to purify raw sugar liquor. The process enabled sugar of very poor quality to be boiled at a low temperature before being filtered through animal charcoal. Boivin and Loiseau had approached a number of British refiners but Henry was the first person to purchase a patent.

The refinery at Love Lane was a success. In its first year in 1872 it melted just over 400 tons of cane from Peru, Mauritius and the East and West Indies. By 1881 output had increased threefold and by 1897 the number of employees had grown from 400 to 642.

When Henry Tate began refining sugar, it was marketed in lump form – large conical loaves that had to be broken up with a hammer. The loaves were easy to transport and store, but not popular with customers. In 1874 an engineer from Cologne, Eugen Langen, created a way of making smaller sugar cubes using a centrifugal device that spun sugar. A year later, Tate, with another sugar refiner, David Martineau, bought the rights to the Langen process. It was a shrewd move on his part; he recognised the value the cubes would have to the consumer. But Henry also realised he would have to take the cubes to them. As a result he relocated the bulk of his business to London, buying three hectares of land in Silvertown, on the banks of the Thames in east London. James Blake, a young Scottish engineer who, at the age of 27, had been given the task of installing new equipment at Love Lane, now had full responsibility for the construction of the refinery. The refinery, which is still there today, began operating in 1878 and in the first year 217 tons of raw sugar were melted and made into sugar cubes. But this rapid expansion took its toll on the Tate family financially and Henry had to take his daughter, Isolina, away from boarding school because he couldn't afford the fees.

However, he kept on searching for new technological solutions to the problem of sugar refining and replaced the Langen process with one invented by a Belgian, Gustav Adant, in 1891. By the end of the century the company was refining over 2,000 tons of sugar a year in London and Liverpool.

Henry Tate grew to be the richest sugar refiner of the time and was renowned for his philanthropy, which was on a massive scale. He provided a library block at Liverpool University costing £25,000; he paid for the building of free libraries at Lambeth and Brixton; and he built and equipped the homeopathic Hahnemann Hospital in Liverpool. He had an acute sense of public service and paid for the building of a meeting place, The Tate Institute, in Silvertown for the two hundred people who lived in the area, for he believed that they had nowhere to meet apart from the public houses. The Institute seated eight hundred and had a reading room and a billiard room. It was, he said, to function in 'an apolitical and non-sectarian environment'. The *Stratford Express* recorded Henry's speech at the opening of the Institute, which was typically modest. It reported that he said that:

> Mr Muir [his managing director] had said many flattering things about him, many more than he deserved and he frankly did not feel that he deserved it for having made the contribution of the Institute since if it had been made possible, it was only thanks to the assistance of his employees. He hoped that the building would help them become better chemists, better electricians, better sugar refiners, better men, whatever the social class to which they belonged.

Henry Tate's greatest gift in terms of money was the National Gallery in London now known as The Tate Britain. A generous patron of the arts, he'd already built a special gallery at his house in Park Hill, Streatham, to exhibit his paintings, including one that has become well known: Millais' *Ophelia*. Each year he held a dinner for the artists he represented and it was said that it ranked second only to the Royal Academy of Art's annual banquet. Henry not only paid for the Tate Gallery, he bequeathed a number of his paintings, including the *Ophelia*, as well as *The North West Passage*, also by Millais; Waterhouse's *Lady of Shallot*, and numerous Landseers, Tademas and Reids. A journalist wrote of him in the *Liverpool Review* in 1891, 'Mr Tate has a big heart and no desire to make it public; he has done his good works effectively but in the greatest stealth.'

Tate as a person was synonymous with sugar – quotations for refined sugar in London generally corresponded to the price of Tate cubes. Henry Tate retired in 1896. He had been twice offered a baronetcy and refused, only yielding when Lord Salisbury, the prime minister, explained that his refusal was a snub to his sovereign. He was finally made a baronet in 1898 but died barely a year later in 1899.

A painting of him in his later life shows a slightly portly gentleman in a suit and black bow tie. He is in a dark mahogany-panelled room in a leather-backed chair, gripping the arms as if slightly uncomfortable with the situation. His beard and short, cropped hair are almost completely white and his dark eyes behind wire-frame glasses scrutinise the painter as if attempting to analyse him. Henry has the demeanour of a strict yet kindly man. He was described by John Hutchinson (publisher of the journal *Notes on the Sugar Industry*) in 1901 as a 'strong, vigorous and kindly personality . . . Even then his speech and appearance denoted that open, honest frankness characteristic of many Lancashire businessmen who have risen from modest beginnings to eminence.' Tom Jones, author of a biography of Henry Tate, said he had lived 'in an age impressive in its greatness at one end and remarkable in its squalor at the other. He had seen the rise of a great industrial middle class of which he himself was a notable member – but he never forgot the deficiencies of the social system of his youth.'

So a genuine, kind, generous man – a philanthropist but shy, reserved and uncharismatic: how then to explain his success? Chalmin comments:

Henry Tate remains an interesting example of the successful Victorian industrialist: as a businessman, he possessed the essential gift of tenacity, an ability to take risks (and above all a knack for analysing situations rapidly), a talent for surrounding himself with the most competent men. He does not strike us as having been a 'leader' nor as having been endowed with any particular charisma or imagination . . . He was not known to harbour clear-cut political opinions and was never seen to take sides, even on sugar problems. A personality which it is difficult to grasp therefore, but one which remains very original among the great figures of Victorian capitalism.

After Henry's retirement, his son William Henry managed the company assisted by his brothers Alfred and Henry Jr. After their deaths, William Henry's two sons, Ernest William and Alfred Herbert (known as Bertie), ran Love Lane. Bertie's claim to fame was his concern for the horses who transported the sugar. The carters waiting to be loaded used to line up in Pall Mall, along which Bertie would walk to the refinery. He would critically examine the horses and their harnesses and, if one looked particularly well kept, he gave the carter a tip. The men vied with each other for his money and his approval. In general Henry Tate's sons and grandsons were competent but undistinguished men.

In spite of the Tates' concerns for welfare – in an era without pensions or paid holiday, they built a surgery and convalescent home – conditions in the refineries were horrendous. 'Just as unmeasured indulgence in sugar is nauseating to the palate, so was the reek of it palling to one's sense of smell. You could taste its clammy sweetness on the lips just as the salt of the sea may be so discovered while the ocean is yet a mile away.' So wrote James Greenwood in *The Wilds of London* of his first visit to a boiling house, a sugar refinery in Whitechapel, London, in 1876. Sugar, he says, in its unrefined state, is foul and black. The walls of the boiling house were coated with a black crust of sugar that hung in stalactites from the ceiling; the sugar refiners, baked in the steamy heat of the place, were semi-naked and caked in sugar, a dirt-encrusted glaze which cracked and split as they moved. 'The heat was sickening and oppressive, and an unctuous steam, thick and foggy, filled the cellar from end to end.' To deal with the heat, the refiners drank beer by the gallon; it was their one perk, but the greatest contributor to their poor health and short lives.

The sugar was melted in giant copper cauldrons before being decanted into moulds. Greenwood added:

A dozen men of the semi-naked sort like those below crawled like frogs over the surface of the sugar moulds, getting foot and hand hold on the edges, some with a sort of engine squirting a transparent liquor into the moulds, and others stirring the thick stuff constantly in the latter with their hands. 'I should imagine that you are not that much addicted to the consumption of sugar,' I remarked to our guide.

It is not surprising that Greenwood's guide had lost his taste for the substance. Conditions were so bad and so dangerous – men were sometimes boiled alive in the sugar vats and occasionally the refineries exploded due to the pressure that built up from the steam used to melt the sugar – that few men would work in the refineries, and so immigrant workers were shipped over from Germany.

One hundred years after Henry Tate was born, the firm he had created would merge with the Lyle corporation, representing a dynasty founded at almost exactly the same time as that of the Tates, and these twin parallel dynasties would come to dominate all others, subsuming them within their own companies.

Abram Lyle III was born in Greenock, Scotland, exactly one year after Henry, in 1820. He was the son of a cooper and fishing-smack owner who made a small fortune but squandered it through his addiction to alcohol. Abram Lyle III was his sixth child and the eldest son; he began work at twelve as an apprentice in a lawyer's office but left when he was fourteen to join his father's cooperage firm. In 1849 his father died, leaving debts of £7,000. This was not the only legacy of his father's: Abram was so against drink after seeing it ruin his father and having 'lived amidst the horrors of a drunkard's household' that he became teetotal. He once told a public gathering that he would rather see his son carried home dead than drunk.

Abram was five foot seven and on the stocky side with red hair and a double chin. He married Mary Park when he was 26 and, although he married again after she died when he was 62, Park was the mother of all his children. Abram took out a bank loan to clear his father's debts and then repaid it by making a success of his father's cooperage business. His venture into sugar refining was carried out with his best friend, John Kerr. These two had started their business dealings as youngsters: they packed and salted herrings for the West Indies when only boys, though the fish unfortunately rotted during the voyage. Lyle and Kerr became sugar refiners in 1865 by founding the Glebe Sugar Refinery along with three other men: James and Walter Grieve and Charles Hunter. Abram managed the new company, which became financially successful almost immediately – within the first year the output of their new refinery doubled, making it the largest in Greenock.

Abram was a competent and shrewd businessman, active in public life, especially in Greenock. He became Provost of Greenock and director of the North British and Mercantile Insurance Company and the Glasgow and South Western Railway. At the suggestion of his wife, he built the Lyle Road as a way of relieving unemployment and to give the townsfolk a pleasant Sunday afternoon stroll. It was dubbed by the Rev. David McRae 'The Heights of Abram'. By the time Abram was 45 he was earning £75,000 a year in present-day terms and during his lifetime gave his sons £230,000 (something over £3.5 million today).

Like many bright men, though, Lyle could be impractical. The Lyles used to have large family dinners, and at one end of the dining room in Oakley, his Greenock home, were two life-size bronze figures holding torches lit by gas. During one of these dinner parties, Abram smelled gas and searched for the leak with a lighted candle. There was a large bang as one of the bronze men exploded and fragments of metal were embedded in the plaster of walls and ceilings. Fortunately, no one was hurt.

Kerr and Abram were well matched: the former intuitive and swift, the latter more cautious. They were inseparable for twenty years, even taking holidays together. But the partnership was cut short when Kerr died in 1872, aged just forty. Abram continued the shipping and the sugar businesses with the help of his sons, who ran the company themselves from 1876. They sold most of their sailing fleet and started again with a small collection of steamships.

In 1880 the Lyles decided they wanted to run their own business and the other partners bought them out of the Glebe. That same year, the Lyles purchased a piece of land in the marshes of Silvertown, in Plaistow, London. It was just over two kilometres away from the Tate refinery by the Thames. However, it wasn't the financial bonanza the Lyles had been expecting. The family's finances were hit hard as the cost of construction went well over budget. Even worse, a massive influx of continental sugar beet knocked the bottom out of the raw sugar market.

The Lyles were forced to sell six cargoes of Java sugar, losing £20,000 per cargo. Their chief creditor, the Bank of Scotland, panicked and demanded a reduction in the loan, but granted them an extension when the Lyle brothers pointed out that there would be no return from an unfinished refinery on derelict land. They finally sold one of their steamships to pay the debt.

The Lyles had no staff with which to run their hard-won refinery. Not trusting Londoners, they engaged three of their second cousins (great-grandsons of Abram I), Alexander, James and John, as managers. Alexander was given the task of finding staff from Greenock and persuading them to move to the south. Just before Christmas 1882, a trainload of men and women arrived at Canning Town station from Scotland. They walked with their families and their possessions through the mud and past the leafless trees to Silvertown.

They found a marshy wasteland with only a few scattered houses at Silvertown village; nowadays the place is a suburb, little different from other poorer parts of London. The train stops in Silvertown itself, right next to what was Henry Tate's factory. The Thames is clogged with huge cargo boats; City Airport is almost next door with planes taxiing down a runway alongside the river; towerblocks, council flats, decrepit pubs and half-demolished corner shops are cheek-by-jowl with new hotels and glass-and-steel office blocks for the businesses many hope will help this deprived section of the docklands regenerate.

The refinery started melting on 10 January 1883, and the first sugar was sold a week later. To begin with even the managers could not afford to buy a newspaper and the men were not always paid. But by the turn of the century the company was highly profitable and the brothers were wealthy men. The refinery at Plaistow initially produced around 400 tons of sugar a week in 1883; this increased to more than 2,000 tons by 1900.

The Lyles' policy was to produce a few types of sugar as cheaply as possible and to depend on one speciality, golden syrup, as the main source of earnings. The syrup, first developed in 1885, was made from a blend of sugar and molasses. In this field the Lyles were technologically ahead of rival manufacturers so, though there were other syrups, the Lyle product maintained its pre-eminence in quality. Its formula is still a jealously guarded secret. When it came to choosing a trademark Abram's religious background influenced his decision. He picked the story of Samson, hence the image on their distinctive green and gold can of the lion killed by Samson and surrounded by bees, with the quotation 'Out of the strong came forth sweetness', from the Old Testament Book of Judges.

Henry Tate and Abram Lyle never met and neither did their sons. Oliver Lyle, Abram's grandson, described how the Tates and Lyles often travelled on the same Great Eastern train from Fenchurch Street station but avoided using the same compartment. Once he found himself walking behind the director of Tate's, so he turned on his heel and headed for the refreshment room, where he dawdled until his chief competitor had boarded the train. Oliver could then make sure he was able to sit in a different compartment. In spite of this, they had an unspoken agreement not to poach each other's speciality: Lyle kept off cubes and Tate did not make syrup. In the 1890s the Lyles heard a rumour that the Tates were contemplating going into syrup production. It's not clear whether the Tates really were going to compete with the Lyles over syrup, but the family responded immediately, buying part of a cube plant. It remained unused and the Tates never did launch into syrup production.

Abram died in 1891 with his sons already running the company. Alexander and Robert worked in Greenock in his shipping business and Abram, Charles, William and John ran the refinery at Plaistow Wharf.

Henry Tate and Abram Lyle III were both remarkable businessmen with a talent for taking important risks at the turning points in their careers. Both had a difficult start in life and both left flourishing companies to their children. Whereas Tate was discreet in his private life but a profound liberal and philanthropist, Lyle was active in public life and participated in the disputes of his times. Both men had numerous offspring – Abram III had ten children, including seven sons, all of whom survived; Henry had five surviving sons and a daughter.

As the nineteenth century came to a close, the Tate Gallery was opened (in 1897) and, at the start of the twentieth century, in 1903, Henry Tate and Sons went public. Since the nineteenth century, in addition to the Tates and the Lyles, another five sugar families had been established: Walker, Kerr, Martineau, Fairrie and MacFie. But, in spite of pressure from these other refiners, not to mention intense competition from imported sugar, by 1914, the three sugar refineries, two belonging to the Tates, one to the Lyles, were producing some 10,000 tons a week of various brands of sugar and syrup.

What could explain these two families survival when other companies were collapsing? They had their problems: the Lyles made many abortive attempts to diversify – they sank a lot of profits into a chocolate factory that failed, for instance; the Tate company quickly began running out of new, young Tates to take over the management. Yet, overall, Chalmin says that they were successful because both families ran their firms well, producing a range of classical sugars at competitive prices, as well as specialising: the Lyles in their luxurious golden syrup; the Tates in their practical cubes.

Ex-Tate & Lyle director Michael Attfield has another theory. He says, 'One of the reasons they were successful was because they were pretty professional about their buying and selling whereas most of the other refiners speculated and went bust.' He adds, 'In my lifetime, I've seen the sugar price at £9 a ton and £600 a ton, so if you make mistakes it doesn't take long to wipe you out.'

The First World War affected both families in the same way, financially and personally. Brothers Oliver and Philip Lyle and their cousin William joined different army regiments – William was killed at the Somme in 1916. Oliver married quickly, in September 1914, by special licence. Both he and his best man were broke so after the ceremony he took a collection from the five guests who were present to pay the parson. He served as a Captain in the Highland Light Infantry and was wounded at the battle of Loos. He was awarded an OBE for his work in the Inventions Department of the Ministry of Munitions.

The war brought huge instability to the businesses. Men left the refineries in their hundreds to join up – many were never seen again. The U-Boat war resulted in the loss of many commodities, including raw sugar on its way to the refineries. The attacks prompted a parliamentary committee under the Earl of Selborne in 1916 to 'consider and report upon the methods of effecting an increase in homegrown supplies having regard to the need for such an increase in the interest of national security'. Although sugar from sugar beet was more expensive to grow and produce than sugar cane, it had the advantage of lower shipping costs, as it could be grown and refined in the UK and Europe. In addition, it became heavily subsidised by the government with the result that Europe would eventually begin to export more sugar than it imported. This was to have a direct effect on the Tates and Lyles right up until the present day: firstly, because the company became involved in this highly lucrative industry; and secondly, once it no longer participated in growing sugar beet, it had to compete against sugar from sugar-beet farmers. This could only be detrimental for Tate & Lyle, who specialised and continue to specialise in refining imported sugar from sugar cane.

Germany and Austria-Hungary had been supplying 66 per cent of British sugar imports. Belgium and France, usually additional sources, were battlefields. So the government took control of sugar refining in 1914, introducing rationing: 8 ounces (about 250 grams) per person per week. In a clever strategic move, the Lyles had anticipated the war and bought huge stocks of raw sugar. The government immediately requisitioned these, though the Lyles made £350,000.

The government also set up a Royal Sugar Commission to dole out supplies to the refiners. Robert Park Lyle was one of the members on the board of the Sugar Commission, heralding the start of a close cooperation between the refiners and the British Government. Through the Commission the government fixed a retail price to avoid high price rises for the consumer on what was considered an essential product, and calculated realistic margins for the retailer, the wholesaler and the refiner. For the first time the refiners were able to work with a guaranteed margin without any risk of fluctuation in prices. Their main concern was to increase production capacity as much as possible. However, the Commission did not fix a price for golden syrup, possibly since it was not seen as an essential product in the same way as sugar. The Lyles therefore were free to put up the price and continued to make vast profits throughout the war.

The First World War marked a turning point in British refining. After periods of uncertainty when the British market experienced the fiercest struggles between foreign producers, five years of relative calm without competition allowed the refiners to both amass a profit and initiate fruitful relations with the government. The immediate post-war crisis was survived thanks to the Sugar Commission. A new period in the history of sugar was about to dawn.

8. SWEET SMELL OF SUCCESS

> Sugar is the favoured child of capitalism.
> Fernando Oritz, *Cuban Counterpoint*, 1947

Emil Fischer's father wanted him to be a businessman like himself. After Emil left school, his father set him to work in his lumber firm. Within a few months, he concluded that his son was too stupid to be a businessman and that he'd better become a student. Fischer, who was an extremely self-confident young man, does not seem to have been damaged by his father's comments. He had always wanted to study physics, but was converted to chemistry by Adolf Baeyer, who became his PhD supervisor. Among other discoveries, he deciphered the structure of glucose and its relation to the rest of the sugar family. As the Tates and the Lyles were growing in size and wealth, Fischer's discovery, relating to the very substance they bought, refined and transported, was a revelation.

Glucose was referred to as 'grape sugar' by the Arabs as long ago as the twelfth century. Andreas Marggraf, who isolated sucrose from sugar beets, describes isolating *eine Art Zucker* [a type of sugar] from raisins in the 1740s and, even then, he realised that this kind of sugar differed from the sucrose obtained from sugar cane. Grape sugar turned out to be identical to the sugar in the urine of diabetics (who cannot store sugar) and was named glucose in 1838 by the French chemist, Jean Baptiste Andre Dumas.

Before Fischer began working on the sugar family, chemists had an idea that glucose was composed of six carbon atoms, twelve hydrogen atoms and six oxygen atoms ($C_6 H_{12} O_6$) but they didn't know what the molecule's structure was. When he was only 23 Fischer discovered a new compound, phenylhydrazine. In 1884, aged 32, he began working with sugars, of which there are many types – glucose, fructose (a sugar found in fruit), lactose (a milk sugar) and maltose (which exists in malt and barley). Sugar itself, sucrose, was at that time still a mystery.

Fischer found that, when he dissolved glucose and fructose in phenylhydrazine, they produced beautiful crystals. He reacted chemicals exhaustively with glucose, fructose and other sugars, in order to determine what their chemical structure was. In addition, he used a polarimeter. By passing light through a sheet of polarised glass and the sugar molecule, he was able to measure the change in angle of the light coming through the other side. Each part of the molecule changes the angle subtly. By doing all this, Fischer showed that the hydrogen and oxygen atoms exist in pairs, attached to five of the carbon atoms, but, on the sixth carbon atom, there are three hydrogen atoms and one oxygen atom.

In 1890 Fischer also demonstrated that glucose exists in two versions, an L-form (for left-handed) and a D-form (for right-handed). The two versions are mirror images of each other, but in practice only the right-handed version is found in nature (which is why glucose is sometimes referred to as dextrose – from the Latin *dexter*, on the right side). Glucose and fructose are each monosaccharides – simple sugars. Fischer thought that sugar itself – sucrose – was composed of a fructose and a glucose molecule, but he didn't know how they were linked or what the structure of sucrose might be. In 1914 in a scientific paper, he wrote, 'We know nothing definite on the mode, how the fructose residue is linked in cane sugar, thus leaving huge room for speculation. I, however, gladly renounce it.'

Fischer was right – glucose and fructose do bond to form sucrose. He also had some other ideas that proved correct. He thought that glucose might not just exist as a linear molecule – he had a hunch that it could also form a chain. We now know that glucose can be looped round in a ring and is a three-dimensional structure – the carbon atoms zigzag up and down from each other and the oxygen and hydrogen groups stick out so that the whole molecule looks like a crazy-shaped chair. Because glucose can open out, it can join on to another molecule, such as fructose, to create a disaccharide like sucrose, or hook up with more molecules to form polysaccharides, such as starch.

However, when an acid, like lemon juice or cream of tartar, is added to sucrose and it is heated, the bonds holding the glucose and fructose together are broken. The mixture becomes thick and treacly and is said to be 'inverted'. This is exactly what the Lyles – without any of Fisher's scientific knowledge – were doing when they created their golden syrup.

During his lifetime Fischer was described by his peers as gentle but authoritarian, 'remarkable', with 'a superhuman capacity for work', and possessing 'phenomenal keenness of mind and flashlike comprehension'. One of his colleagues wrote of him, 'The princely man who gathered about him most of the doctoral students and other young investigators overshadowed everyone in greatness, spirit and scientific insight.' In 1902, at the age of fifty, Fischer was awarded the Nobel laureate for his ground-breaking work in chemistry. Sadly he was plagued with ill health and tragedy. His wife, Agnes Gerlach, died of meningitis after they'd only been married for seven years and two of his three sons died young: one caught typhoid fever on military service in Romania and the other committed suicide. Fischer himself was poisoned by the very chemical he had discovered, phenylhydrazine. On 11 July 1919, in severe pain, he was diagnosed with intestinal carcinoma. On 14 July he spent most of the morning with his lawyer making a will, in which he established a scientific foundation to support young German chemists. He mailed his last two papers, both on sugars, and then took his own life.

Before Fischer dissected the component parts of sugar, people had long appreciated sugar's remarkable properties, even though they did not understand why it behaved the way it did. In the sixteenth century baked desserts first appeared, such as sweet custard in pastry, and these puddings owed their taste and appearance to further properties of sugar: caramelisation and the Maillard reaction.

Caramelisation occurs when sugar melts at 160°C, forming a clear liquid. As it gets hotter, sucrose breaks down into glucose and fructose and water is evaporated out of the solution. The glucose and fructose then break down even further and recombine with other molecules in the pudding. As this chemical reaction progresses, the liquid changes from clear to yellow to deep caramel. A whole new range of molecules are formed – it is a process we barely understand but the end result is that new flavours are created and, for example, the delicious golden coating adds the sweet crunch to *crème brûlée*.

The Maillard reaction also explains why sugar turns brown, but the process is slightly different to caramelisation. It's named after Louis-Camille Maillard who was interested in the biochemistry of living cells and discovered how amino acids – proteins – react with sugar inside the cell. Ironically, Maillard never worked with food and yet his name has become inextricably linked with the kitchen. Maillard browning happens when sugar is combined with a protein – from eggs, for instance, in a cake. The protein bonds with the sugar and, as the cake is heated, water evaporates. The surface, in particular, dehydrates and this new protein-sugar complex turns brown and crisp.

During the nineteenth century ordinary people had grown accustomed to a plentiful supply of sugar, bread and jam, biscuits and sweet tea, Victoria sponge and penny ices. Hannah Glasse's Chinese temples constructed from sugar pastes had fallen out of fashion, but a whole new era opened up for sugar.

As the Tates and the Lyles grew and expanded their business during the nineteenth century, there was a concomittant rise in the number of factories solely dedicated to producing sweets. Few of these sweet factories remain – Stockleys of Accrington in Lancashire, Brays in South Glamorgan and Buchanan's of Greenock are the survivors – but the sweets they made still fill some of us with nostalgia: rhubarb and custard, sour plums, fairy satin cushions, clove rock and cinder toffee.

Malcolm Stockley learned the art of sweet-making as an apprentice in a sweet factory but the First World War temporarily intervened. On his return he set up on his own, creating two-tone pear drops, sour apples and sasparilla in the front room of his terraced house in Accrington in 1918. The factory has expanded somewhat and continues to make 'boilings', as hard, boiled sweets are known, in nearby Oswaldtwistle Mill.

The main trick in sweet-making is to prevent large crystals from forming that feel gritty on the tongue. At Stockleys the sugar is dissolved in water to make a thick syrup and heated to a specific temperature, usually 112–116°C, before being poured into a traditional copper vessel that heats the sweet solution in a vacuum. Heating sugar in a vacuum means the syrup reaches its setting point at a lower temperature. As the solution is warmed, it becomes more concentrated because water evaporates, but the hotter it gets, the more soluble the sugar becomes and the more sugar that the solution can hold. It's ready when a teaspoon of the mixture dropped in cold water will form a soft ball.

The syrup needs to cool rapidly to prevent crystals from forming, so it is poured on to flat trays that have cold water circulating beneath them. The syrup is clear and it is at this point that flavours and colours are poured in and stirred through – red dye, for instance, seeping through the syrup like a Rorschach ink blot. Gradually the syrup starts to solidify until one of the workers can pick up the dense slab and fold it in on itself, the hot innards of this fat sugar square bursting open in thick, red gashes. It oozes back out to fill the tray until, after several foldings, it remains in a heavy wedge, the surface a dull, opaque burgundy. It is rolled into a portly sausage and fed into a batch roller, a machine that squeezes it between a number of cogs, increasingly pushing it into a narrower and narrower rope. Finally, it's chopped into lozenges that continue to coil round a drum, semi-attached like a vast, deep-purple millipede, before being swept into a giant basket to be taken away for packaging.

Stockleys sell fudge too – the old traditionals like rum'n'raisin and chocolate, as well as newer varieties, such as coffee. Fudge is made in a similar way to boiled sweets but is cooled slowly. To prevent crystals from forming, 'interfering' agents are added right at the beginning of the process – like fat from butter, for instance. Once its set point has been reached, fudge is cooled to 40°C – it's a highly unstable mixture since it contains more sugar than it should have at such a low temperature. It is beaten to seed the mixture with tiny crystals – you can tell when this happens because the mixture will lose its glossy sheen and becomes so thick it barely flows. At this point it's poured into trays and left to cool.

It seems nothing short of miraculous that the sap from a grass can, with the addition of heat and a juggling of ingredients, be turned into chocolate fudge, cough drops, cake, custard or lemon curd.

As you walk up the road from the train station there is a strong odour unlike anything you will have smelled before. It might be from the adjacent sandwich filling company, but it's how I imagine toasted cat food might smell. The Thames Refinery in Silvertown owned by Tate & Lyle is the largest in the world. Sprawling over 13.36 hectares between the road and the railway line on one side and the Thames on the other, it still has 120-year-old buildings dating back to when it was first built in Henry Tate's day.

Back at the start of the twentieth century, the Tates and the Lyles were still both family-run firms. The Lyle factory was down the road from Silvertown. The two were to merge in 1921, but the company remained paternalistic, controlled as it was by family members, almost until the present day. It continued to maintain close links with the government and was characterised by traditional ideas. Yet it became the largest and most powerful sugar company in the world. Though its position as market leader has been surpassed in the last few years, today Tate & Lyle is one of only three of the original one hundred FTSE companies that have survived.

The refinery, which used to be the largest in the world, processes just over one million tons a year, at a rate of twenty-three thousand tons a week. When I was there, being shown round by Community Relations Manager Michael Grier, the *Handy Esperance* was unloading at the dock. The 33,000-ton ship had a hold literally filled with raw sugar. Bulldozers worked inside the ship, shovelling the sugar into a radio-controlled bucket elevator. It was a bleak, grey day and the crew had to work fast; with a tidal drop of 5.5m, the ship would not be able to remain for long at the dock while fully loaded. The sugar is stored in a warehouse, stacked in piles over 6,000 tons deep. The floor is black and sticky with sugar and there's a faint, sweet caramel smell.

This raw sugar, the colour of pale sand, is usually 97–98.5 per cent pure; the impurities are mainly proteins and minerals from the sugar-cane plant itself. The whole long, costly, highly technical process of refining sugar until it is the purest substance known takes place to remove that 2–3 per cent of impurities and turn raw sugar into almost 100 per cent pure white crystals of sucrose. It is a far cry from the Jamaica train or oxen crushing cane in a giant pestle and mortar.

What happens first is that the raw sugar is heated with impure syrup to create magma, a brown, bubbling porridge, which is poured into a battery of centrifugal machines. Each one contains perforated baskets that spin at 1,050rpm for two minutes. First they revolve in one direction, then the other. As they stop, warm vapour rises from the spun sugar, like the hot breath of a horse. At this point the sugar is 99.2 per cent pure. The remaining impurities are locked within the sugar crystals themselves. The sugar is dissolved once more and this time milk of lime (calcium hydroxide) and carbon dioxide gas is bubbled through the syrup as it is swirled through vessels two storeys high. The gas reacts with the lime to form chalk (calcium carbonate). The chalk attracts impurities such as waxes, gums and resins and is fil-tered out by pumping the liquid sugar through steel blue-green machines that look like alien musical organs; a panel of glass columns above the machines resembles the pipes and the liquid is forced through a series of cotton filters – the bellows of the organ as it were. Impurities collect against the cloth. After further processing, this cake-like substance is removed and sold to farmers as a soil improver.

The remaining insoluble impurities are then removed by ion exchange – the liquid sugar flows through banks of vessels containing resin beads, the size of ball bearings. The negatively charged resin removes coloured molecules from the sugar solution. The sugar is now 99.9 per cent pure and is given one final polish by passing it through a column of carbon.

The final purification process is designed to crystallise the sugar out of the solution. The liquid is fed into huge steel vessels called vacuum pans, embedded with round portholes capable of holding more than sixty tons of liquor. Because the syrup burns easily, it's important to crystallise the sugar and get rid of the water but keep the temperature low. Steam is used to heat the pans, which are pressurised to create a vacuum so that the sugar need not be heated to such a high temperature. This is the time when real skill and expertise are required, as the operator has to determine when the sugar is ready to form crystals of exactly the right size. It used to take nearly three years of training to become an engineer; now it takes six months.

The process was simplified by fitting television cameras to the portholes. The images are then relayed to a small room next to the vacuum pans and magnified five times. As the crystals dance and swirl through the solution, the picture gradually darkens until the screen turns black, clogged with sugar crystals.

At this point, the damp sugar is spun in granulators, or 'grannies'. The heat within the crystal itself is enough to help evaporate the water that remains, clinging to the face of the crystals. The floor round the grannies is covered in a fine white powder and the air is laced with hot, tropical aromas. The drying sugar can behave as a liquid and would be liable to explode if the company did not take appropriate safety measures.

Although the sugar now appears dry, there is residual moisture within the crystals themselves, which can cause the sugar to clump together, so it is given a final blast of hot air. Grier recalls that once, when he was working as an engineer in the Greenock refinery in 1970, there was a sharp frost. The fans inside the silo where the sugar was stored broke and the result was the world's largest sugar lump. The men had to dig the 600-ton sugar cube out by hand. It took them a month. There were far fewer health and safety rules for businesses in those days.

The final stage is to package the sugar and this part is almost completely automated. The place is eerily devoid of people. It takes half a second to wrap one kilo of sugar; the smallest package is a four-gram sachet of sugar. The packets are printed on a wide roll of paper, which is sliced from the main roll, cut into its component parts, glued together, filled with sugar, and sealed in an elegant robotic ballet. As we watched the production line, Grier told me of a dinner-party trick: put a teaspoon of brown sugar in your coffee, he said, holding the spoon just below the surface of your drink. Then lift it out. The sugar will now be white. Although you can buy raw sugar, brown sugar is often refined sugar with caramel flavourings and brown colourings sprayed over the white crystals.

In 1918 Ernest Tate approached Charles and Robert Lyle with a proposal for a merger. The Tates were anxious about their future; although their melt was twice that of the Lyles, they were making a lower profit overall. But the real reason for Ernest's proposition was the absence of young Tates to take over the management of the firm. The Lyles, on the other hand, had no shortage of young men – two of Abram's sons and four of his grandsons ran the business. A merger presented a perfect and profitable partnership. In Liverpool the Tates were well placed for the market in the north and in London each firm maintained its speciality, cubes and golden syrup. But, by merging into a single firm, they would control half of Britain's sugar refining. This would have been an attractive proposal given that, with the end of the war, would come the end of the Sugar Commission and, as a result, foreign competition would resume.

Although negotiations for a merger began in autumn 1918, an agreement was not reached until early 1921. No one knows why it took so long as the points of difference were minor. One problem was how to harmonise the directors' salaries. The Tates' salaries were four times higher than the Lyles and the same went to a lesser extent for the rate of return on preferential shares. In the end, both shares and salaries were equalised. Since Tate was a public company, they had to buy Lyle and then exchange shares. But eventually the two firms joined forces on a fifty–fifty basis.

The merger took place on 27 February; the Sugar Commission had stopped functioning the day before. The *International Sugar Journal* ran the following story:

> The most significant event in the history of the British refineries in the past few years is without a doubt the current absorption

by Messrs. Henry Tate and Sons Ltd. of the refining business of Messrs. Abram Lyle and Sons. A certain community of interest increased by the restrictions of the war has undoubtedly led these two large London-based companies to merge and we can expect substantial economies of administration, purchase of raw materials and distribution of finished products . . . However, the creation of something resembling a refining trust in this country must be condemned both from the point of view of the raw sugar producer and from that of the end consumer and therefore trends in this direction should be followed closely.

For the new Tate & Lyle Ltd, the negotiations and comments within the sugar industry were worth it: the firms no longer competed for raw sugar; together they managed to share technical expertise and even have a reasonable profit on granulated sugar. However, at the beginning of the 1920s there was a drop in British sugar consumption and foreign refined sugar resumed fierce competition. By 1926 foreign sugar amounted to 40 per cent of all British consumption. The main competitors for British refiners were the US, Czechoslovakia, the Netherlands and Canada. The situation deteriorated further in 1924 with the creation of a government subsidy for British beet sugar.

During the First World War beet fields had turned into battle-grounds across Europe and beet production had plummeted. World sugar production, though, had doubled, as sugar-cane plantations tripled between 1910 and 1930. At the time the Institute of Agronomical Research at the University of Oxford carried out a study showing that it would be possible to cultivate beet in the west of England. Because the world market was fluctuating rapidly, the government solved the problem in two ways. Firstly, they encouraged beet cultivation on British soil and a subsidy was paid to farmers so that they could obtain a decent price for their sugar beet. Secondly, supplies from the Empire were increased. By the time the Second World War had broken out, the country had achieved sugar independence.

That same year, 1924, Arthur Guinness, the financial secretary to the treasury, expressed surprise at the refiners' lack of haste in becoming involved in British beet production. What he didn't realise was that Tate & Lyle were already in the process of creating a sugar factory at Bury St Edmunds for beet sugar partly using Hungarian capital and technical expertise. Oliver Lyle said, 'We had been told that the

Hungarians were practically the best beet-sugar manufacturers and that explains why we associated with a Hungarian group.' Oliver and his brother Philip went to Hungary to look at beet factories and he describes how in one factory he stepped back to get a better view of the evaporators, tripped and fell into a vat of raw beet juice. The edge was only a half a metre above the floor and there was no rail. Fortunately, with the help of the Hungarians, he was pulled out unharmed. He said:

> I was taken to the laboratory and swabbed down. I could not change my suit as I only had the one with me and it soon grained out in a kind of sucro-textile suit of armour. Hungary, while we were there, was in the throes of wild inflation. We stayed a night in a great castle where a beautiful young countess lived in the grand manner and was served by five butlers, one of whom acted as my valet and did his best at remelting my suit. When we left I gave him what was intended to be five shillings. At par exchange it was £40,000.

Tate & Lyle formed another four companies under the Bury umbrella, which, by that time, accounted for a third of British production of sugar beet. Oliver said, 'After one or two years, as soon as one plant began to run well and the personnel acquired experience, profits would rise.' Later, in 1936, the British Sugar Corporation was created, a quasi-governmental body controlling beet sugar. Tate & Lyle owned the majority of shares and, when the British Sugar Corporation requisitioned the Bury group, received almost £2 million in cash in compensation.

In 1928 the Conservatives, headed by Stanley Baldwin, were in power and Lord Leonard Lyle, Abram's nephew, was a Conservative MP and close friend of Winston Churchill, then the Chancellor of the Exchequer. The government amended their 1928 budget to include changes in the scale of customs duties that favoured raw cane over refined white sugar. The concession would clearly help Tate & Lyle Ltd. The *International Sugar Journal* became concerned, commenting that 'there is very little sugar being produced or refined in this country which is not under the control of Tate & Lyle'. The *Public Ledger* expressed astonishment at this amendment, describing it as 'granted exclusively to Messrs. Tate & Lyle' and commenting that it must have come from 'higher up' than the director of the Customs and Excise Department. Chalmin concludes:

> The days of poverty were a thing of the past. Imports of refined sugar dropped spectacularly, raw sugar imports increased and exports of refined sugar even reached levels beyond the wildest dreams of the most optimistic a few years before . . . the refiner received more than he had initially paid (the duty was paid on raw sugar but reimbursed on refined sugar).

As a consquence, Tate & Lyle reduced prices to the consumer over the following years. In the meantime, though, they also embarked on a shopping spree, buying half the Walker company of Greenock, the third-largest British refining centre, enabling them to control 50 per cent of the capital of the Glebe Refining Company; the Fairries' refinery and Macfie and Sons of Liverpool. Tate & Lyle now refined three-quarters of all sugar processed in Britain.

During its first twenty years Tate & Lyle developed a strong company character. In 1924 the management was a mixture of members of the Tate & Lyle families and of employees who had worked their way up to the board of directors. Whereas in other businesses the descendants of the founders were being pushed aside by a new race of trained managers, Tate & Lyle took the opposite path. The Lyles and the Tates had become products of the British establishment, attending public schools and studying at Oxford or Cambridge. Although science graduates were hired during the 1930s, it wasn't until after the Second World War that any were appointed to the board. The company's management was a collegiate one; each director had specific responsibilities. Overall the firm was very paternalistic. Priority was given to employing the children of persons already working for the company – in parallel with the Tate and the Lyle dynasties, legions of employees emerged with up to three generations working for the firm at the same time.

In the 1930s Tate & Lyle's major innovation was the creation of a brand: with the aid of packaging machinery from Germany they could print their logo on to paper packets to encourage customer loyalty. Previously consumers had bought sugar that was weighed for them in the grocer's. Now they could not only buy sugar in packets 'untouched by hand', but they were also specifically targeted as consumers and encouraged to generate a loyalty towards the brand. It was a major innovation, perhaps even more innovative than the packaging machinery itself. It enabled the name Tate & Lyle to become synonymous with sugar in Britain; in other countries, Tate & Lyle's subsidiaries fulfilled the same function.

The company continued to expand – into transport, creating Eastbury Transport Ltd, for road transportation, and Silvertown Services Ltd, which controlled a fleet of tugs and barges. At the same time it grew into a global firm by buying plantations and sugar factories in Canada (Redpath), Jamaica and Trinidad. The company instantly became the largest Jamaican producer, producing a third of all sugar, and the second-largest Trinidadian producer. Political instability and difficult working relations plagued Jamaican relations with the company, but Trinidad, to begin with, proved to be a gold mine.

Tate & Lyle's West Indies Sugar Company (WISCO) was founded in 1937 in Jamaica. The first few years were difficult; the British West Indies was in the throes of a decade-long economic crisis. Poverty and unemployment – 14,000 were unemployed in Jamaica in 1938 – were prevalent; often those who were able to find work found that conditions and pay were bad. Wage demonstrations were held and, in May 1938, Jamaica's new trade union movement was born. Alexander Bustamente, half-Irish, half-Asian-Indian, was a born orator looking for a cause. Although he was arrested, public unrest was so great that he was released and went on to found the Bustamante Industrial Trade Union (BITU) and helped create the People's National Party, led by his cousin, Norman Manley.

The directors of Tate & Lyle responded by improving housing, hygiene and increasing salaries. New cottages were built to replace the slums that the workers had lived in; medical and sports facilities were provided and plots of land were made available to rent in an effort to improve the workers' diet. Manley took over the BITU in the 1940s and, with his more systematic and less flamboyant approach, he was able to negotiate greater pay increases and a week's paid holiday.

Because of their paternalistic outlook and swift response, WISCO experienced a remarkably profitable decade in Jamaica. However, the death knell for WISCO was falling sugar prices combined with Tate & Lyle's wish to modernise the most labour-intensive part of the sugarcane operation, the cutting of the cane by hand. This would mean significant job losses, and was blocked by the government. A committee set up to examine Tate & Lyle's involvement, the Mordecai Commission, accepted that mechanisation was necessary, but wanted it phased in gradually. Without it, Jamaica's productivity for Tate & Lyle dropped until it was no longer worth their continued investment in the country.

There is, of course, no suggestion that Tate & Lyle had ever been involved with slavery. It is unfortunate, therefore, that they chose to buy plantations complete with the 'Big House' that were a legacy of slave-owning days and which they purchased from foreign companies, such as Barclays Bank, that still controlled much of the sugar industry in these Caribbean islands. Sir Saxon Tate, who later became the chairman of Tate & Lyle, joined the company when he was just twenty in 1951. At the age of 23, he was sent out to Jamaica and Trinidad and he comments:

They were very much under the thumb of the white man, very much colonies run by an old-boy network. The management . . . were archetypal colonists – people in pith helmets slapping their boots. When I said the management style was unacceptable I got belted for it. I was told it was not my place, at the age of 23, to say those things.

The Mordecai Commission said of the problems in the sugar industry in Jamaica, 'The tone adopted lends considerable credence to the general complaint by the trade unions that managers are determined to maintain the intolerant master and slave relationship of the colonial era.' Michael Attfield, the first non-family member of Tate & Lyle to become a director, disagrees that the company's failure had anything to do with the legacy of slavery and people's long and bitter memories: 'Growing sugar cane was hard work and . . . it's a lovely sunny place, why work?'

Chalmin writes:

By 1965 in Jamaica one could no longer speak of a plantation economy. All the external vestiges of it were still there: the big house, the cane fields stretching as far as the eye could see . . . but the sugar industry had ceased to be the motor of the country to become instead one of its increasingly lame ducks, a ship which was gradually letting in water and which its captains, tired of plugging the holes, would gradually begin to abandon.

In 1937 there were also a number of riots in Trinidad; one resulted in twelve deaths and the looting of a sugar factory. Like Jamaica, the riots led to the creation of trade unions, but Trinidadian independence happened more quickly, less violently and with less opposition to the company. In Trinidad Tate & Lyle's company, Caroni, had a 29,000-hectare estate sprawled over a 45km stretch of the island producing 55 per cent of the total sugar-cane output. Caroni was responsible for grinding 90 per cent of the island's sugar cane and buying cane from 80 per cent of the smaller planters. Its production capacity exceeded 200,000 tons of sugar per year. During the sugar-cane harvest it employed 15,000 people but in total 150,000 were dependent on the company. It profits grew from £500,000 in 1953 to £1.8 million in 1963. Dr Jean-Claude Gaicottino, whose PhD thesis was on sugar cane in Trinidad, said that Caroni 'constitutes an agro-industrial empire whose power could exert an influence on the newly independent state'.

In 1956 the People's National Movement (PNM) was founded in Trinidad; when the country achieved independence in August 1962, Oxford academic and historian Dr Eric Williams became the prime minister and remained in power until his death in 1981. In spite of his views on Western capitalists, Williams declared in 1966 at the PNM convention: 'The best policy in the national interest is to produce sugar as efficiently as possible whilst settling unemployed workers on state lands where they can plant food crops.' Caroni took over two more plantations, including the only one left that had been controlled by the Trinidadians, Woodford Lodge. According to Chalmin, Caroni functioned in a harmonious environment in Trinidad – there were relatively untroubled social relations, cordial relations with the government and the firm experienced external growth. From 1946 to 1965, Caroni was profitable for Tate & Lyle. It was not to last and, after Trinidad, Tate & Lyle's forays as landowner in other countries were cut back.

Meanwhile, back at home in the 1930s, as prospects for peace deteriorated, the British Government turned its attention to solving the problem of how to keep the country supplied with food during another wartime. On the advice of Sir William Beverage (head of the Ministry of Food during the Second World War) a Food Department was created to ensure that total control could be exercised over the entire British food system once war broke out. Work began on a product-by-product basis and from July 1937 ration cards were printed. Stockpiling was instituted, but was hampered in the case of sugar due to a poor beet harvest in 1938–39 and a rise in the world price of sugar cane. War was declared on 3 September 1939 and a week later all sugar stocks in the UK came under the control of the Sugar Division of the Ministry of Food. Sugar was rationed to ¾ of a pound (300 grams) per person per week (as opposed to 2 pounds (800 grams) before the war); industry was allowed 60 per cent of its usual requirements. On 27 May 1940 the ration was reduced to 8 ounces (200 grams) per person per week and industrial consumption was cut by half.

Once again the company was safeguarded by a Lyle, Philip Lyle. Tate & Lyle converted much of its Thames refinery to other uses: they dehydrated potatoes, cabbages and carrots for the Ministry of Food, while the fitting shops produced gun parts, spares and shuttering for gun emplacements. Due to men being called up to fight, from 1941 women had to be employed and from 1942 they filled half the refinery jobs (although they only earned three-quarters of the men's salaries). It was a female employee, Doris Martineau, at the Thames plant who suggested launching a collection in the firm to buy a bomber for the RAF. Everyone contributed and in three months £20,247 had been raised. Tate & Lyle purchased a Whitley Bomber called the *Golden Lion*, which was donated to the RAF.

At all refineries the drill for a crash shutdown was honed until it took only seven minutes from full melt to closure without blowing off telltale steam. On the night of 9 May 1941, however, the company offices at Mincing Lane in East London were burned out – almost all their documents and papers were destroyed. The company chiefs accommodated staff made homeless in the bombing raids (up to 300 people slept at each refinery). They laid on a fleet of 24 coaches to bring people into work and sent company vans to help salvage furniture from employees' damaged homes.

In spite of these problems Tate & Lyle's profits never dropped below £1 million and exceeded those made in the 1930s. Rationing continued in Britain until 1953, with sugar remaining at 8 ounces per person per week. But by the end of the war Tate & Lyle's turnover totalled £78.5 million, the largest ever recorded.

In 1945 the Labour Party had become increasingly outspoken in its advocacy of a 'Socialist Commonwealth' and in the postwar election of that year they swept to power. In 1942 the party platform announced, 'Only collective ownership will guarantee the pre-eminence of national interest over private need.' So, in January 1947 the National Coal Board was created, in 1948 the British Transport Commission and the British Electricity Authority, and in 1949 the Gas Council. In the 'Labour Believes in Britain' pamphlet presented at the 1949 annual conference on 13 April the following paragraph appeared:

> Sugar refining is also controlled by a state-protected private monopoly, which has enabled large profits to be made for private shareholders. One concern dominates the industry; it also has large interests in a few Colonial territories that the Labour Government is pledged to develop. The sugar industry is vital both in war and peace. Labour intends to transfer to public ownership all the sugar manufacturing and trading concerns.

Leonard Lyle disputed the claim that Tate & Lyle refined most of the sugar in the UK; his list of other producers showed that Tate & Lyle owned only 52.8 per cent of the market share. Chalmin, however, calculates that not only did the company virtually control cane sugar, they also still owned part of beet sugar's British Sugar Corporation's capital and one of their directors was on the board, which meant they had first-hand knowledge of their rival industry. Chalmin puts the real figure at 88 per cent: 'All in all in 1949 ... Tate & Lyle genuinely controlled – directly or though its influence – the British sugar economy.'

But of all the large business owners threatened with nationalisation, Tate & Lyle, under Lord Leonard Lyle, was to launch the most well-publicised counterattack. Lyle's response was to create a character to attack Labour, but the success of caricaturist St John Cooper's creation took even the company aback. Mr Cube – a talking sugar cube – waving his white fists in the air and shouting, 'Tate not State', was stamped on the back of every sugar packet. A total of ten million anti-nationalisation slogans were printed a week. Other slogans were: 'Sugar may be nationalized. Mr Cube says, Only the State will make my price jump!' and 'Mr Cube says Oh dear dear! Dearer!'

A memo from Mr Cube aimed at the British trade unions at their annual conference at Blackpool was sent out in autumn 1949. Thirteen million copies were printed, describing British sugar as the best and the cheapest in the world. Mr Cube appeared as a cut-out toy – a million copies were distributed to children. Posters were pasted to the firm's delivery vans. Employees were not immune – the *Tate & Lyle Times* was launched, presenting the anti-nationalisation arguments. A grand sports day was held with the election of Miss Tate & Lyle, 'the sweetest thing in sugar', and the winner, Maureen Wigger, became a 'spokeswoman' for the campaign. The firm selected speakers to present their case to clubs for the elderly, for women, the Young Conservatives and Liberals, the Rotary and Lions Clubs.

A survey in November 1949 showed that just over half of the population knew that sugar had been threatened with nationalisation and, of those that knew, only 13 per cent were in favour. Leonard Lyle claimed that their campaign was apolitical. In his book, *Mr Cube's Fight against Nationalisation*, published by Tate & Lyle, he reiterated speeches he made at the time.

We never had any wish to become associated with any political party in this fight. Nor have we become so. I make this point because since the threat came upon us from one particular political quarter we have inevitably been thought of as the opponents of one party and the allies of another. This is not the case. We have been, throughout, concerned only to get sugar refining removed from the list of industries destined for nationalisation.

However, the nationalisation of sugar – particularly of Tate & Lyle – became one of the major issues in the election campaign. In the 1950 Labour Party election manifesto it was announced that vital branches of industry were becoming monopolies that should be blocked by nationalisation – Tate & Lyle were high on the list of dangerous companies.

Ex-director Colin Lyle, Philip Lyle's son and Abram's great-grandson, says of the time, 'I can remember when I joined the company and we were threatened with nationalisation we certainly financed ... a major PR campaign to do down the government's policy.' The last element of the Tate & Lyle campaign was a petition against nationalisation. When it was completed it had just over one million signatures.

The campaign worked. In the 1950 general election the Labour Party kept its majority but lost seats and was in a weaker position than it had been before. The King's Speech at the opening session of Parliament declared, 'A law will be proposed to make the legislation on the beet sugar industry permanent and to transfer the shares of the BSC which are at present not held by the Treasury to public ownership.' Leader of the Opposition Winston Churchill was not fooled and the next day said in the Commons, 'The proposal about beet sugar, which was no doubt intended to keep alive the nationalisation issue, somehow seemed, while letting off both barrels, not to have hit Tate & Lyle. That was the one they were aiming at but they shot at a pigeon and hit a crow.'

There was another general election in October 1951, which Labour lost. This defeat of the Labour Party buried the nationalisation question for good.

Though the issue was resolved, the battle drove Tate & Lyle to avoid any similar problems in the future by going global. The company directed its efforts towards sugar production and refining in other countries, particularly the West Indies and Africa. They shipped sugar and offered their expertise to help other companies manufacture machinery and build sugar plants throughout the world.

In 1965 they began to become a truly multinational company. United Molasses (UM) was a wealthy firm founded by a Dane, Michael Kielberg. It was an international molasses company and, in addition, traded in alcohol, grain and controlled a 267,220-ton tanker fleet and a 90,428-ton fleet of ships and cargo vessels. W.A. Meneight worked for the company for fifty years and wrote *A History of the United Molasses Co. Ltd.* He says that, in response to general despondency about profits, a glossy brochure was published to publicise United Molasses, which included a map of the world with a large number of dots showing where United shipped or received molasses. Although rich, however, United Molasses was suffering from uncharacteristically low profits. Kielberg and his contemporaries had died and, according to Chalmin, it was a ship without a captain.

On 11 March 1965 Tate & Lyle put in a takeover bid. Colin Lyle opposed it:

> I was concerned with biting off more than we could chew and growing too much, and then becoming a slave to City opinion. What worried me was that we knew nothing of their business. The fact that we produced molasses as a by-product totally disguised the fact that we knew nothing about trading in molasses . . . The only man who could deal with it was Michael Attfield, who was a brilliant commercial man and trader.

After the deal had gone through, Attfield was sent out to UM for six months. He says, 'I never worked so hard in all my life. It was slightly difficult, as United Molasses didn't want to be taken over. I was like Daniel in the lions' den, so to speak.' He adds, 'The Tates were very slow. They should have marched in with their boots and taken much tighter control from the word go – we were far too gentlemanly about it.'

Although the takeover had been contested by the UM directors, Meneight concluded that, 'United was taken over by a Sugar Daddy and left intact . . . it must be said that many in retrospect have vouchsafed the opinion that the takeover was beneficial as a whole.' The acquisition increased Tate & Lyle's fixed assets by £22 million.

It generally took no more than five to seven years for a Lyle or a Tate to be called to the board. Charles Lyle was only 24 when he joined in 1929; Colin Lyle was 27 in 1954; and Saxon Tate was only 25 in 1956. Saxon's mentor was Sir Oliver Lyle. He says:

> When I joined the company Oliver said, 'You have to do all the jobs in the refinery. When you can boil a pan of sugar as well as the leading pan man, you'll know the business.' Which is a very curious way to teach someone business. What in actual fact happened was I learned an awful lot about people working on the shop floor and factory life and made a lot of friends, that I've never, ever regretted – but learn to run a business?!

Colin Lyle thought that he would probably be the last generation in the family business and still believes he is correct. There is, however, one member of the Tate family still working at Tate & Lyle. Colin says:

If you came in as a trainee shift manager, there was a discreet arrangement whereby the managers and the directors asked the foremen, 'How's he doing? Is he going to make the grade?' [Colin makes a sign as if having his throat slit if the answer to the question was negative] . . . I sacked two Lyles in my time on that basis. If you're going to justify a system of nepotism and make it credible to all the employees, the only people who could get to the top were the ones they themselves, via the foremen, voted for.

When Colin became a director, 47 per cent of the employees were relatives of the original Tate & Lyle workforce. The family itself theoretically had little control, owning only 8 per cent of the shares, but they dominated the board and controlled the decision-making.

In 1955 Michael Attfield and a colleague were the first two graduates who were not family members to be given management training. Now in his 70s, Attfield, who asks me to pass him 'the pension' when he wants a spoonful of sugar in his coffee, says that, as fresh, young 21-year-olds, 'We were known by all the senior management as Burgess and MacLean [two notorious Soviet agents who had infiltrated the British establishment and fled to Russia in 1951], because we were regarded as spies from head office.'

In spite of the merger having taken place nearly forty years before, the Tates and the Lyles still continued to run their business as two parallel dynasties: Tate employees rarely spoke to Lyle employees; Plaistow remained a Lyle refinery, Silvertown a Tate one; and beneath it all there was a shimmering undercurrent of antagonism. The firm remained family run and paternalistic well into the 1980s. As Colin says, 'You may call the firm Tate & Lyle. Many of us enjoyed [calling it] Tate versus Lyle.'

It would take several decades before the company would change in response to the unfolding political situations in Africa, the nationalist movement in the West Indies, Great Britain's entry into what was then the European Economic Community, and the decreasing number of new Tates and Lyles born into the family until finally, by the 1980s, no family members would remain on the board. In the meantime, Chalmin notes that Tate & Lyle oversaw sugar from its agricultural roots right through to the finished product on the consumers' table: from the sugar-cane fields through to refining, packaging and transport. Even the staff for their new subsidiaries were hand-picked from those who already worked for the company and were, no doubt, related to generations of employees who had laboured in this familial firm.

In 1963 Tate & Lyle began discussions with the government of Belize, which, at the time, was still a British colony. It was a site with virgin land and easy access to America, a market Tate & Lyle hoped to infiltrate even further now that Cuba was enduring a US embargo. Belize Sugar Industries Ltd was formed and Tate & Lyle began a vast programme of investment – £7 million over three years. In seven short years the country's entire economy was dominated by sugar; it accounted for three-quarters of its total exports and employed 40 per cent of the labour force.

Belize was only granted small sugar quotas by Great Britain and the US so Tate & Lyle had to run at a loss until 1971. After its experiences in Jamaica and Trinidad, the firm decided that it was dangerous to own plantations and, with the agreement of the government, sold off its lands to independent farmers. They kept just 404.7 hectares as an experimental farm and created a service to disseminate technical knowledge.

In 1969 Tate & Lyle expanded into South Africa, taking control of the fourth-largest sugar group in the country, the Illovo Sugar Company. Sugar-cane cultivation had been introduced in South Africa at the end of the nineteenth century and was initially concentrated in the coastal region of Natal, before spreading to the interior, to Zululand and south of the Transvaal. The South African industry was extremely tightly knit; three companies, which had extensive financial links with one another, owned all the sugar refineries.

By 1977 Illovo had become a prosperous firm. But in 1977 the British independent television network ATV criticised the working conditions at Illovo. Tate & Lyle protested, putting together a dossier that claimed certain interviews had been faked. But from the British viewpoint it looked as if the company was condoning apartheid and its social consequences. Despite its profitability, Tate & Lyle decided to sell Illovo in August 1977.

Throughout the 1970s Tate & Lyle gradually abandoned or reduced its traditional activities such as refining in Great Britain and sugar production. In 1965 the company owned plantations in Jamaica, Trinidad, Zambia and Belize and controlled sugar refining and distribution in Nigeria and Rhodesia (now Zimbabwe). They expanded into new activities such as sugar trading under Michael Attfield's lead. Trading results quadrupled and shipping profits doubled. In 1975 the results were even more spectacular: trading alone accounted for 61 per cent of the entire group's profits, while refining in Great Britain and sugar production also showed remarkable results. These performances were repeated in 1976 when Tate & Lyle's profit before tax was a record £52 million, the majority from trading. The firm then increased its investments, mainly in shipping.

Tate & Lyle would experience an even more meteoric rise under director Neil Shaw, previously the director of their Canadian subsidiary Redpath, as he bought up most of the major American refineries and the beet sugar industry. In just fifteen years, from 1965 to 1980, the company evolved more than in the whole preceding century. By the end of the 1970s it had truly become a multinational. But from 1976 to 1977 the group's long-term indebtedness also almost doubled from £58 million to £112 million.

Although they were expanding worldwide, back in Britain Tate & Lyle found the 1970s difficult. In late 1973 the coal miners' strike crippled the country and a three-day working week was enforced on industry to conserve energy. During the winter industrial action led to more working days being lost through strikes than at any other time since 1926. Against this background, Tate & Lyle had to face the impact that Britain joining the European Economic Community would have on its UK sugar-refining business.

Other European countries resisted importing raw cane sugar, but Britain insisted that cane supplies should be maintained from developing Commonwealth countries. The high-profile debate that followed eventually resulted in the 1975 Sugar Protocol of the Lomé Convention, which was signed on 1 February. It decreed that African, Caribbean and Pacific (ACP) countries were to be given quotas to supply sugar for EEC countries at a guaranteed 'intervention' price. Extremely efficient sugar-producing countries, such as Australia, were denied access, as were developing countries not in the ACP protocol. This meant an immediate reduction of over 500,000 tons of raw sugar entering the UK. At the same time, beet sugar was given massive subsidies; the consequence of this was that excess sugar from non-ACP countries and from beet was dumped on the world market, further pushing down the global price. According to Saxon Tate, this meant that developing countries 'were systematically shot in the foot by Brussels. Six years out of seven the world price is practically below the cost of everybody else's production.'

The immediate impact, however, was that there was insufficient sugar being imported to keep the sugar refineries in Britain, including Tate & Lyle's last competitor Manbré and Garton, going.

Alexandre Manbré, born in 1825, came from Valenciennes in the old, northern French province of Hainaut and was the son of a farmer; like many farmers in those days he was also a brewer. Sometime between 1855 and 1857 Manbré came to England. He applied for two patents in London in 1858 'for the manufacturing of colouring spirits from the sugar of potatoes, known as glucose' and 'for the extracting of saccharine matter in malt ... for the purpose of brewing and distilling'. On 1 January 1864 Glucose Sugar and Colouring Ltd was registered as a company – at approximately the same time as Henry Tate and Abram Lyle were starting out. The firm produced glucose, starch, gum and colouring matter. Manbré's was first bought out in 1919 by Albert Berry and was then taken over in 1927 by Sir Richard Garton, becoming Manbré and Garton Ltd. Garton retired that year and there was no further involvement of any of the families that had founded and worked in the firm. But other sugar families, the Martineaus and Kerrs, bought into the company.

By 1970 Manbré and Garton were refining over 400,000 tons of sugar a year, making them the third-largest British sugar company after Tate & Lyle and the British Sugar Corporation. In 1976 they took over Fowler Ltd, a small cane-sugar refinery on the banks of the River Lea in Blackwall, which produced a range of products including West Indies' Treacle.

Manbré and Garton had been given 30 per cent of the sugar-cane quota and Tate & Lyle the other 70 per cent. Saxon, who was chairman at the time, pushed forward the idea of a takeover bid. He says:

It was perfectly clear that the raw sugar supplies were not going to be sufficient for the refineries then in business. The only way you make money out of a sugar refinery is to run at one hundred and ten per cent of its capacity seven days a week. Manbré wasn't in a position to do that. There was really simply no way we were going to agree with the others whose refineries were going to be closed down.

Michael Attfield adds that Tate & Lyle had to protect their core business, especially as beet sugar was continuing to expand and was so heavily protected by European subsidies. In July 1976 the company made its offer.

There was a considerable risk that the proposal would be brought before the Monopolies Commission. Certainly Manbré and Garton and many sugar users tried to demonstrate that, if Tate & Lyle controlled the entire refining sector, including approximately half of Britain's glucose and 40 per cent of its starch production, Tate & Lyle would also have a monopolistic position in the sweeteners markets. However, on 9 September the decision was taken by the government not to refer the takeover bid to the Monopolies Commission.

Manbré and Garton cost a total of £48 million – £20 million too much according to Attfield. But in one fell swoop Tate & Lyle had bought up their entire domestic competition. However, it was, says Attfield, a big and expensive mistake, partly because Tate & Lyle knew nothing of Manbré and Garton's starch business. He says, 'We should just have left them alone and they would have withered on the vine, so to speak.' Tate & Lyle committed themselves to extensive and expensive refurbishment of Manbré and Garton's Battersea refinery, which they could not afford. In Attfield's opinion, 'We spent too much money, we should just have shut it down straight away.' By 1979, in an abrupt about-turn, the refineries were closed and 4,600 workers were left without work.

Despite the closure of Manbré and Garton, by the end of the 1970s and into the early 1980s, Tate & Lyle were in serious trouble. The core finance of their firm was based on trading, which the city considered to be unstable: shares had fallen dramatically. They had spent heavily; in addition, sugar prices had risen dramatically and beet sugar continued to be a serious challenge. As a multinational firm they had difficulties due to the political situation in their African companies as well as the Caribbean ones, while American refineries had been bought at a high price but had produced little sugar.

Tate & Lyle took the decision to withdraw from most of their global concerns, retaining only Zimbabwe. Back in Britain, the Liverpool refinery begun by Henry Tate was closed. The shipping and transport subsidiaries were sold. Their research laboratories, in Kent and later in Reading, were scrapped.

There still remained, however, the company's inherent problem, riven right through it like writing within a stick of rock: in spite of their global status, Tate & Lyle was still an old-fashioned, family-run firm. Although this way of running the company had worked well for many years, in the 1970s, Saxon Tate felt that changes needed to be made. He brought in Lord George Jellicoe, an ex-banker and politician, and between the two of them they 'cleaned up'. Colin Lyle disagreed with this; he felt Jellicoe was a politician, not a manager and that his appointment was 'in contrast to the firm's tradition of mateyness, team spirit and morality'.

'It was frightful, it really was,' says Saxon, 'I emptied out thirty-five top managers and some of them were at an awkward age and I know that some of them were terribly resentful because Tate & Lyle didn't *do* that sort of thing.'

Finally, to add insult to injury for Tate & Lyle, for the first time in history, British sugar consumption was shrinking. There were two reasons for this. The first was the development of a new type of sweetener that could replace sugar in processed foods. In 1966 two Japanese scientists discovered an enzyme (glucose isomerase) capable of converting glucose into fructose. By converting cheap corn starch into a glucose-rich liquid, 55 per cent of which is then turned into fructose using this enzyme, high fructose corn syrup (HFCS) could be made. In the 1960s, while sugar prices were low, little attention was paid to this new invention but, as sugar prices rose, in the 1970s it became increasingly popular. When Coca-Cola announced that it was going to replace the sugar in all their soft drinks with HFCS, the market began to rise steadily. Today HFCS is the sole calorific sweetener added to soft drinks in America and in 2003 the United States exported $90 million worth of HFCS.

The other problem for sugar producers the world over during this period, and which is becoming an increasing threat to the industry, is the realisation that excess sugar can be detrimental to our health.

9. SWEET NOTHINGS

What is the most meaningless phrase in food marketing? My personal vote would be, '92 per cent fat free' (or any other per cent, come to that). Anyone who thinks this statistic says anything either useful or good about a food product is at best gullible, at worst an idiot (and quite likely, a gullible idiot). Of course, the implication of the 'fat free' slogan is that the food so labelled can be eaten in large quantities without any adverse effect on your weight or your waistline. This has to be the biggest, nay fattest lie the bow-tie-and-big-glasses brigade has ever come up with – not a bad accolade for a profession whose sole business it is to tell porkies to the public.

Hugh Fearnley-Whittingstall, *Observer Food Magazine*, 2004

In the 1930s, a research dentist from Cleveland, Ohio, Dr Weston A. Price, travelled the world, recording his travels in his book, *Nutrition and Physical Degeneration: A Comparison of Primitive and Modern Diets and Their Effects*, lavishly illustrated with hundreds of photographs. His conclusion, recorded in horrifying detail in area after area, was simple. People who live in so-called primitive conditions had excellent teeth and wonderful general health. They ate natural, unrefined food. As soon as refined, sugared foods were imported as a result of contact with 'civilisation', physical degeneration began in a way that was observable within a single generation.

'Let us go to the ignorant savage, consider his way of eating and be wise,' Harvard Professor Ernest Hooton had said in 1937 in *Apes, Men, and Morons*. 'Let us cease pretending that toothbrushes and toothpaste are any more important than shoe brushes and shoe polish. It is store food that has given us store teeth.'

Dental caries or tooth decay was virtually unheard of in early human history. When the skulls and teeth of early humans and traditional hunter-gatherers are examined, there are no signs of modern tooth decay. Dental caries first appeared in the teeth of the rich; it took longer for the diet of the poor to change and it wasn't until the latter half of the nineteenth century that sugar became cheap enough for the poor to eat. By the First World War, tooth rot was so bad that some men were refused entry to the army because they didn't have enough teeth to chew their rations. Rationing immensely improved dental health during both World Wars, but the nation's health deteriorated again after the end of the Second World War.

Tooth decay is caused by eating sugar and refined carbohydrates, which are broken down to simple sugars. As soon as sugar enters the mouth, bacteria in plaque (*Streptococcus mutans*) starts metabolising it and producing acid. The acid not only causes decay itself, but it corrodes tooth enamel, allowing further decay to take place in the softer tissue beneath the enamel. A study that examined animals' teeth at London Zoo showed that the animals didn't have caries, nor did they even have the *Streptococcus mutans* bacteria. It was only in fruit-eating monkeys that they found this particular mouth bacteria, which is not naturally found at high levels.

The more sugar one eats, the more *S. mutans* builds up. These bacteria flourish in the acidic environment created by eating sugar and have a sticky coat enabling them to cling to teeth – interestingly, studies have shown that rats only have *S. mutans* on their teeth when they feed on sugar cane. We inherit our bacteria from our parents. Parents kiss their children and taste their food as they feed them, literally infecting their children with these decay-causing bacteria. They even inadvertently encourage them to have a sweet tooth by adding sugar to food to apparently make it nicer or buying sweet products because they themselves prefer them. As a result, the ecology of the mouth changes and becomes less habitable for the more benign forms of bacteria we used to have.

Eating less sugar, even abstaining for only three days, seriously decreases the amount of *S. mutans* in the mouth. However, it is important to note that it is primarily the *frequency* of sugar consumption that is highly correlated with dental caries, rather than the *amount* of sugar one consumes.

Tooth decay is the one health hazard related to sugar that some in the industry have admitted to. Saxon Tate says, 'If you get a concentrated carbohydrate solution that is acid round the teeth, it's going to have a very bad effect on teeth. The cola drinks are absolute murder because they're mostly sugar, which is the carbohydrate, and the other ingredients are acid, so you get this frightful situation in the mouth.' He quickly adds, 'Sugar won't damage your teeth if you look after them, but, if you drink beverages that are acid and have a lot of sugar in them and don't clean your teeth, then it will damage your teeth.'

According to other experts, however, it's usually too late to prevent damage by brushing after eating or drinking. Aubrey Sheiham, a professor of dental health at the Department of Epidemiology and Public Health at University College London, says, 'It's one of the biggest myths – brushing does not help reduce tooth decay.' Within just three minutes of eating sugar the level of acidity increases to the point where minerals leach out of tooth enamel, leading Sheiham to say that you would have to be a sprinter to get to the bathroom and clean your teeth before the damage done. He adds that bacteria are tucked into the gum between the teeth and so it is physically impossible to reach them either by brushing or by flossing. In fact, it's better to brush beforehand as this helps remove a bit of plaque, which reduces the number of bacteria to some extent and hence the amount of acid they produce. It is saliva, which is neutral – neither acid nor alkaline – and buffers the acidity and remineralises teeth, that prevents your teeth dissolving if you eat sugar and don't brush. However, this process doesn't start for about twenty minutes.

'Teeth are the hardest structure in the human body and yet we're dissolving this substance. If this happened to any other part of the body – if sugar burned great holes in your throat – it would be banned immediately,' says Sheiham. It would be as if we went to Mars and found that the Martians had dreadful foot rot. Their toes and toenails kept dropping off and they had to visit other Martians who would make toe replacements and silver and gold toenails. Incredulous humans would ask, 'What is the matter with your feet?' And the Martians would reply, 'Oh, we like eating Mars bars but they make our toes disintegrate.' Sheiham concludes, 'But people accept tooth decay due to eating sugar.' Dentists, he says, would not exist were it not for sugar.

The sugar industry is well aware of the problem of caries: the new attack, when it came, was from quite another quarter entirely and an unexpected one. John Yudkin was a professor in the human nutrition department at Queen Elizabeth College, part of the University of London, in the 1960s. He was a self-confessed sugar junkie: 'If only people knew how many pounds of milk chocolate and liquorice allsorts and cakes I used to tuck away each week! At a rough guess, I would say that my total sugar consumption must have been not less than ten ounces [250g] a day, probably nearer to fifteen [375g].'

But Yudkin had a problem: he suffered from dyspepsia due to a duodenal ulcer diagnosed a quarter of a century earlier. As middle age approached, Yudkin noticed that he could no longer eat those pounds of chocolates and cakes and remain lean: he began to spread round the middle. He started to try out various diets and discovered that by reducing his intake of carbohydrates he not only lost weight, but his ulcer stopped being painful. Carbohydrates – starch-containing foods like pasta, bread, potatoes and cake – break down into simple sugars, but can also contain added sugar themselves. Yudkin's theory was that, because sugar is acidic, it irritates the lining of the gut wall. He corroborated his hypothesis by examining the gut walls in rats and people after they'd eaten sugar. He also persuaded his patients to try his diet. It helped alleviate symptoms in 70 per cent of his patients and many of them also lost weight. Yudkin's no-carbohydrate diet was thus one of the forerunners of the Atkins diet. He published his new diet book in 1968.

Shortly afterwards, Yudkin realised that, when he had advised people to avoid carbohydrates, he had not made a distinction between starchy carbohydrates, such as potatoes, and sugar. It was sugar, he began to think, not carbohydrates per se, that was the enemy. In 1972 Yudkin published *Pure, White and Deadly: The Problem of Sugar*. In his book he took the sugar industries, both in the UK and in the United States, to task. Sugar is a pure and natural substance, which gives us energy. This is true but, as Yudkin pointed out, it depends on the definition. Firstly, sugar is the purest food known – it is literally 100 per cent sucrose. But in food, purity is not necessarily valuable. If a substance is pure sugar, it does not contain any vitamins or minerals, and these are vital. Secondly, sugar is indeed natural – it comes from the sugar-cane plant – but it has been refined to such an extent that it is hardly a natural substance any more. Finally, there is the concept of sugar as an energy-giving substance. Yudkin said:

What people really mean when they say that sugar gives them energy is simply that it is a potential source of the energy needed for the process of living. It is there when you need it, in the same sort of way as the petrol that you put into your car is in the tank, ready to be burned when you want the car engine to go. Just putting another gallon or two in the tank does not, of itself, make the car go any faster, or make it any more energetic. And taking another spoonful of sugar does not, of itself, make you jump out of your chair and rush to mow the lawn.

Sugar provides calories – four per gram – but they are empty calories, since sugar does not contain any nutrients. So, for example, if the average man eats a healthy and nutritious diet amounting to 2,500 calories a day, and then replaces 500 calories' worth of his normal food with 500 calories of sugar, he should not put on any more weight – he's still eating the same total amount of calories – but now he's eating 500 calories less of foods that contain essential vitamins and minerals. We need simple sugar – glucose – to survive and we can obtain this from our diets. Refined, processed sugar is not necessary, since for thousands of years of human evolution we did not have any. Indeed, our bodies are not designed to eat it – we still have the physiology of cave men and women.

The point is that we *like* eating refined sugar. All of us love the taste of sweet foods. In our past this would have driven us to find the sweetest, ripest fruit; now it leads us to pick the choicest chocolate gateau, strawberry cheesecake or *crème brûlée*.

What was different was Yudkin's litany of disorders and diseases caused by sugar and his major and surprising hypothesis – that sugar increased the risk of having a heart attack. Up until this point, a strong correlation had been made between saturated fats – fats from dairy products and meat – that increase cholesterol levels, causing arteries to harden, leading to heart disease. But Yudkin claimed that there was, in fact, a better correlation between the consumption of sugar and coronary mortality in the twenty countries studied. Furthermore, he showed that patients who were suffering from coronary or vascular disease ate twice as much sugar as healthy patients.

The sugar industry, of course, must have been dismayed by Yudkin's findings, but none of the ex-directors I interviewed had actually ever read his book.

There were flaws in Yudkin's research. Some of the experiments he carried out were over a very short period of time (only up to three weeks), involving a small number of people – mainly young men who were forced to eat rather large amounts of sugar. Perhaps the reasoning was that only nineteen-year-old boys would consent to consuming half a kilo of sugar a day.

However, Yudkin's rather flamboyant style – 'If only a small fraction of what is already known about sugar were revealed in relation to any other material used as a food additive, that material would promptly be banned' – and his call to the government to ban sugar entirely fuelled a health scare. The sales of packets of sugar dropped.

Today, most nutritionists and the majority of long-term dietary studies show that heart disease *is* caused by eating too much saturated fat, not by eating sugar. But Yudkin was not entirely wrong. He was very careful to explain that he did not think sugar was always the *direct* cause of disease. He wrote:

> None of the diseases that I shall be talking about are caused by sugar in the same sort of way that heat causes ice to melt. People differ in their susceptibility to disease, so that even in identical conditions – supposing you could produce them – one man might have a heart attack and another might not. This susceptibility seems to a large extent to be inherited, so one may say that your chances of getting a coronary are less if your parents, grandparents, uncles and aunts have mostly lived to a ripe old age without having the disease.

He also added that a high intake of sugar was not the only 'cause' of heart failure. Coronary mortality also correlates with obesity, a sedentary lifestyle and possessing a television. As he said, 'The best relationship of all existed between the rise in the number of reported coronary deaths in the UK and the rise in the number of radio and television sets.' In other words, if you eat too much, don't exercise, have a sedentary lifestyle and smoke, you will be at risk from heart disease. These symptoms – too much fat and sugar and a reduction in activity – are associated with affluent countries, the ones in which people are more likely to die of a heart attack, as well as in wealthy members of developing countries.

In addition, Yudkin pointed out that cholesterol from saturated fats is not the only precursor of coronary disease – fatty acids in the blood, known as triglycerides, also rise and can increase the risk of cardiovascular disease. Many experts now agree that triglycerides can be produced by eating too much sugar. According to the American Heart

Association's 2002 statement on sugar and cardiovascular disease, a diet high in sugar (about a fifth of total calories being regarded as 'high') will increase triglyceride concentration in the blood. A study published in 2003 by the *American Journal of Clinical Nutrition* has also shown that sugar is to blame; both sucrose and fructose increase the levels of triglycerides by 60 per cent more than starchy carbohydrates do. In other words, sugar can contribute to the risk of a heart attack.

But can sugar turn into fat?

Sugar, or sucrose, is a disaccharide, made up of a fructose and a glucose molecule joined together. When you eat glucose the level of glucose in the blood, also known as blood sugar, rises. Some glucose is immediately used – even at rest the body burns calories, and more will be burned if you are exercising. The rise in blood sugar triggers the pancreas to release insulin, which converts glucose into glycogen, a complex carbohydrate that is then stored in the liver and in the muscle cells. The average adult has about a teaspoon of sugar in their blood, which gives them enough energy for fifteen minutes. Thereafter, they replenish their blood sugar from stores in the liver. The liver can hold up to 1,800 calories' worth of glycogen – about a day's supply. This is accepted fact. What follows is not.

One theory about what happens next is that, once the liver's stores are full, any more glucose that is eaten is converted into fatty acids and these, if not used immediately for energy, are stored in fat cells around the body. There is much debate in the medical literature over the issue of whether glucose is actually turned via glycogen into fat and stored in adipose (fatty) tissue.

Levels of obesity have risen alarmingly: in the UK one in five children are now obese or overweight. This is an increase of 50 per cent in only twenty years. Obesity is calculated using the Body Mass Index (BMI), which is a measure of body fat based on a person's height and weight. It is calculated by dividing weight in kilos by height in metres squared. A person with a BMI between 25 and 29.9 is considered overweight; over 30 is obese. In the UK just over one in five adults are clinically obese – 21 per cent of men and 23.5 per cent of women; a further 47 per cent of men and 33 per cent of women are overweight. This means that over half the adult population is either overweight or obese. The situation in the US is even worse: one in three adults is obese and half of all children. What is alarming is that this trend is increasing and has sped up considerably over the past decade.

Obesity-related diseases cost the National Health Service half a billion pounds a year. But is it really sugar that is to blame?

By 2001 global consumption of sugar and other sweeteners had risen to 159.5 million tons, 2.5 times higher than it was in 1961. According to the International Confectionery Association, based in Brussels, the UK eats more sweets and chocolate than any other country in Europe – 23 per cent of the European market, amounting to 14 kilos per head.

In 1996, in response to the government's recommendations on nutrition, which were heavily criticised by the Sugar Bureau, the Bureau itself published a paper which it presented to the Department of Health, entitled 'Nutrition and Health Aspects of Sugar Consumption: A Case for Reconsideration of the Scientific Basis of Aspects of Current UK Government Advice Relating to Sugar'. It featured an appendix by Professor Andrew Prentice, now at the London School of Tropical Medicine and Hygiene. Prentice examined a number of studies that correlated increased weight gain with diet. He concluded that the increase in sugar intake was not the likely cause of rising levels of obesity in the UK; like many experts during the 1980s and 1990s, he thought fat was the culprit. Current thinking at the Lipid Metabolism group at Oxford University, headed by Professor Keith Frayn, is in agreement: in overweight people, the majority of fat deposited in their fat cells comes directly from fat in their diets, not from sugar.

There is, of course, no denying that eating too much fat will make you fat. Firstly, fat contains more calories than any other food source – nine per gram, almost twice as much as a carbohydrate. Secondly, the body only processes fat after it has dealt with alcohol, protein and carbohydrates, which means that fats are more likely to be stored, rather than used. Finally, eating fat does not make you feel as full as protein or carbohydrates do. Our bodies seem to have an inability to register that we have eaten fat (and thus a large number of calories) and so, our appetites unimpaired, we continue to eat more than may be necessary. However, there are problems with the studies cited by Prentice. The scientists involved were all searching for a correlation between fat and obesity, not the correlation between sugar intake and obesity. Unless these two issues are separated and the question of whether sugar also contributes to obesity asked, it is difficult to reach an accurate conclusion.

In 1977 Patrick Holford met two extraordinary nutritionists who, quite literally, changed his life. Over a bowl of salad and some soya sausages, Brian and Celia Wright explained to him how most disease was the result of sub-optimum nutrition. Holford says:

> I found this hard to swallow, but, being an adventurous spirit, asked them to devise me a diet. There I was, a university student studying psychology, eating a wheat-free, virtually vegetarian diet with masses of fruit and vegetables, and taking a handful of supplements shipped from America, since they were not available in Britain at that time. It was a far cry from the usual fish and chips and a pint of bitter! My colleagues, friends and family thought I was crazy. But I persisted.
>
> Within two months I lost fourteen pounds [7 kg] in weight, which has never returned; my skin, which had resembled a lunar landscape, cleared up; my regular migraines virtually vanished; but most noticeable of all was the extra energy. I no longer needed so much sleep, my mind was much sharper and my body full of vitality.

Inspired, Holford himself became a nutritionist, eventually founding the Institute for Optimum Nutrition in London in 1984, where he is still a director, and writing a number of books that encapsulate his ideas on a healthy diet, including *The Optimum Nutrition Bible* and *The 30 Day Fat Burner*. 'Sugar,' he says clearly, 'is bad for you, and it makes you fat.'

Holford's argument rests on a condition known as disglycaemia – losing the balance of sugar in the blood. When you eat sugar, you get a rush of insulin, which swiftly removes sugar from the blood and, as a result, you have low blood-sugar levels and a slump in energy. Low blood-sugar levels are associated with tiredness, an inability to concentrate, irritability, nervousness, depression, sweating, headaches and digestive problems. As blood-sugar levels regulate your appetite, you will swiftly feel hungry again, even though you have just eaten.

This is quite the opposite of the claim sugar companies have made – that sugar, because it gives you a hit of instant energy, is actually slimming. Back in the 1960s, one of the advertisements ran:

Willpower fans, the search is over!
And guess where it's at? In sugar!
Sugar works faster than any other food to turn your appetite
down, turn energy up. Spoil your appetite with sugar, and you
could come up with willpower.
Sugar – only 18 calories per teaspoon, and it's all energy.

Not much has changed. The Sugar Bureau is still trying to persuade
people of the 'slimming' benefits of sugar. A statement on their website
read:

Research shows that people can continue to enjoy some sugary
foods and still lose weight. In fact, eating sugar may increase
their chances of success – by including some low-fat sugary
foods such as arctic roll, jaffa cakes, sorbets, fruit yoghurt, rice
pudding, jelly beans and currant buns in their diet.

However, according to Holford, three out of ten people have
trouble keeping their blood-sugar levels even. The result is that over
the years they grow fat and lethargic. What he believes happens is that
the more frequently you raise your blood-sugar levels, the more
insulin you produce, and the more insulin that is produced, the more
sugar is stored as fat. Over time, you will become insensitive to insulin
and become insulin resistant. Insulin resistance in adults can lead to
Type II diabetes, as well as Syndrome X, which is the name for a cluster
of risk factors for heart disease associated with insulin resistance.
These factors include hypertriglyceridemia (high levels of fat in the
blood), low HDL-cholesterol (the 'good' kind of cholesterol), hyperin-
sulinemia (high blood insulin), hyperglycaemia (high blood glucose),
and hypertension (high blood pressure).

However, while John Yudkin also discusses this effect in his
book, overall there is little evidence that someone of normal weight
who is not insulin resistant to begin with and who then eats large
amounts of sugar will become insulin resistant. Most research in
this area indicates that those who become insulin resistant and who
go on to develop Type II diabetes do so as a result of becoming
overweight. Carrying a large amount of excess body weight
somehow prevents cells from responding effectively to insulin. This
can, to a certain extent, be reversed by losing weight.

Nevertheless, a study published in 2003 by the Institut National Agronomique in Paris seems to support Holford to some extent. It showed that, in rats, sugar alone – not fat or protein or being obese to start with – was responsible for increasing both insulin resistance and obesity.

The situation is further complicated because not everyone reacts to sugar in the same way. There is a sixfold difference in how individuals respond to insulin: some of this variability is due to genetic differences; and 25–35 per cent of the variability is related to being overweight (though a quarter of non-obese middle-aged people are thought to be insulin resistant). Other differences occur because those who exercise will release less insulin in response to sugar and people who normally eat high-fat diets will overproduce insulin when they eat sugar.

But what about the idea that sugar itself can actually be converted into fat and therefore indirectly contribute to obesity?

Dr Toni Steer, a nutritionist from the Medical Research Council in Cambridge, says, 'Our viewpoint is that over the last twenty years there's been a big focus on the role of dietary fat in obesity, which has led to the creation of low-fat biscuits and snacks. This reinforced the idea that fat makes you fat and everyone forgot about the role of carbohydrates and sugars.'

Nutritionists are just beginning to examine carbohydrates again and are starting to realise that not all carbohydrates are equal in terms of health effects. Sugar, as we know, is a carbohydrate, but refined carbohydrates, such as white bread, have already been partially broken down into simpler starches and therefore quickly become converted to glucose in the body. These carbohydrates release energy rapidly but the boost is followed by a sudden slump in exactly the same way that eating refined sugar leads to an energy crash. A portion of instant mashed potato, for example, can cause an almost identical blood-glucose rush to that produced by pure sugar. These foods – refined carbohydrates and starchy vegetables, like potatoes – are known to have a high glycemic index (GI). The definition of GI put forward by Australia's Sydney University Glycemic Index Research Service is:

The glycemic index is a measure of the power of foods (or specifically the carbohydrate in a food) to raise blood sugar (glucose) levels after being eaten.

The Sydney group, run by Professor Jannette Brand-Miller, use glucose as the standard measure – they assign glucose a GI of 100. Other foods are tested by giving ten or more healthy people a set portion and then measuring their blood glucose over the next two hours. These results are then plotted on a graph and the area under their two-hour blood-glucose response, called the AUC (Area Under the Curve), is measured. These same people then consume the equivalent amount of glucose and their blood glucose is also measured. The GI value of the test food is calculated by dividing the AUC for the test food by their AUC for glucose, then taking the average GI value for the ten participants.

The Sydney group are conducting a number of studies, one of which also seems to confirm the idea that there is a variability in insulin sensitivity in the population. Brand-Miller and her colleagues have shown that Asian men naturally have a greater level of insulin resistance than white men of the same weight even when given the same amount of a high GI food.

The field of GI science is not as established as some branches of science – Brand-Miller's group have only been going for a decade – yet the research not only suggests that eating high-GI foods can contribute to obesity but that a low-fat diet, according to Brand-Miller, does not always lead to significant weight loss and usually any weight lost is quickly regained. Brand-Miller says, 'Our hypothesis is that a high-carbohydrate diet based on foods that promote a high glycemic index alters appetite and energy partitioning in a way that is conducive to gaining body fat.'

Substances that release glucose slowly and do not give a high blood-glucose level are known as low-GI foods – examples of these are: sweet potatoes (boiled, not roasted), millet, porridge, wholemeal pasta and pulses. The reason why a low-GI diet works is because, firstly, when you eat low-GI foods there is a low insulin response so your blood-sugar levels don't crash. Second, low-GI foods increase satiety as they release glucose into the blood slowly, so you don't feel the need to eat again straight away. A decrease of 50 per cent in GI, for instance from a food that has a GI of 75 to one of 50, makes you feel 50 per cent more full. This is because low-GI foods stay in the gut longer and are absorbed more slowly, so receptors in the intestines are stimulated for longer, sending chemical messages to the brain's satiety centre to say that you are still full. Finally, low-GI foods can promote fat oxidation instead of fat storage. Long-term animal studies confirm this, showing that a high-GI diet increases weight gain and promotes fat storage: rats on a high-GI diet put on 16 per cent more weight, 40 per cent of which was fat, and had twice as much around their internal organs as rats on a low-GI diet. Even more worryingly, the rats on the high-GI foods showed an alteration in their enzymes, which decreased their ability to burn fat.

In a study of pregnant women, those who were on a high-GI diet put on more weight than those who were on a low-GI diet (19.7 kilos compared to 11.8 kilos). Teenage boys who ate a high-GI breakfast ate 53 per cent more food than boys who had a low-GI meal, and were less able to oxidise the fat that they had eaten. Other research showed that, for several hours after eating a high-GI meal, fat oxidation was suppressed whether the volunteer was resting or exercising.

Thus, as Andrew Prentice says, although little sugar is actually converted to fat, the net result when you eat sugar and high-GI foods is that fat oxidation is impaired, excess fat is stored and fat already in storage will not be used up. Therefore, eating sugar and/or a high-GI diet can make you fat.

Sugar can, however, also be directly turned into fat in some cases: a recent study by the Department of Biochemistry and Nutrition at the Scottish Agricultural College, Ayr, shows that women, whether lean or obese, will convert sugar to fat if they eat 50 per cent more than they normally would.

What is interesting is that, although the science of GI foods is disputed by some, professors of epidemiology and nutrition at the Harvard School of Public Health, Walter Willett and Meir Stampfer, have created a new food pyramid that incorporates these principles. In 1992 the US Department of Agriculture released the Food Guide Pyramid, which was designed to help the American public eat a healthy diet. Willett and Stampfer's alternative pyramid distinguishes between healthy and unhealthy carbohydrates; they suggest people eat those that do not swiftly raise blood sugar. The researchers found that people who followed their guidelines had a lower risk of major chronic disease, and reduced their risk of cardiovascular disease by 30 per cent for women and 40 per cent for men. As Patrick Holford says, 'I have worked clinically with literally thousands of patients and put them on a low-GI diet to help them lose weight. There is no question that it works.'

Andrew Prentice says that he would have written his report for the Sugar Bureau differently now because of changing trends. Since the report on sugar and obesity was published in 1996, he has been alarmed at a worrying increase in the consumption of soft drinks, which has risen dramatically over the last decade and has increased by 400 per cent in thirty years. In the latest National Diet and Nutrition Survey, it seems that over a quarter of sugar consumed by four- to eighteen-year-olds comes from soft drinks. In adults fizzy, sweet drinks alone accounted for 12 per cent of their sugar intake, while sweetened drinks – including alcohol and fruit juice – make up 40 per cent of their total sugar intake. In the US, sugar intake in soft drinks drunk by teenagers has tripled in the last thirty years. People are now drinking more than a hundred extra calories a day from sweetened beverages.

Recent research is beginning to show, however, that sugar in sweetened drinks may prevent your appetite from functioning properly. For instance, in one study published in the *American Journal of Clinical Nutrition*, a number of overweight volunteers either had sugar-rich drinks or ones with artificial sweeteners. The ones drinking the sugary drinks put on weight over ten weeks – an average of 1.3 kilos, almost of all of which was fat (as opposed to lean muscle) – while the group drinking the artificially sweetened drinks lost around a kilo. But the reason the volunteers put on excess weight was not just because they'd had more calories from the drink, but because they also tended to eat more – and generally more sweet things. Almost a third of their energy intake came directly from sugar and, as a consequence, their blood pressure also increased.

A second study gave volunteers a sugary drink or a snack of the same calorific value. They were then allowed to help themselves from a buffet. The people who'd had the drink ate more than the ones who'd had the snack. This may not sound surprising but the calorie intakes in both groups had been the same. Whereas the people who had the snack regulated their food consumption, those who had had the drink were unable to. As Toni Steer says, studies like these show that 'Sugar in the form of soft drinks simply impairs appetite control. If you continue to eat your normal diet and have a couple of cans of soft drink, it's easy to see that you'll get into a positive energy balance – you'll be eating more calories than you burn – and then you'll put on weight.'

A few years ago a group of researchers at the Department of Nutrition, University of California, Davis, started to look into the reasons why obesity has risen so alarmingly over the last two to three decades. Their studies reveal that, since high-fructose corn syrup (HFCS) has been increasingly added to drinks instead of sucrose and many people in the developed world are drinking more sweet beverages, the amount of fructose people consume has risen markedly. In America the average person consumes more than 38 kilos of fructose a year. This works out at almost 400 calories a day from fructose alone. Just two cans of a typical soft drink will supply 200 calories' worth of fructose – more than 10 per cent of the energy requirements for the average woman. One of the authors of the study, Dr Peter Havel, says, 'Fructose consumption now makes up a significant proportion of intake in the American diet, and an increase in fructose consumption has coincided with an increase in the prevalence of obesity over the past two decades. It is therefore prudent to ask whether current fructose intakes could contribute to weight gain.'

Compared with glucose, fructose does not trigger a rapid increase in insulin nor does it increase the levels of a hormone called leptin. So, although fructose does not trigger the blood-sugar crash that happens when insulin drops, too much fructose is a problem because both insulin and leptin regulate hunger over the long term, meaning that they set the tone for your appetite. Thus, as a result of eating a lot of fructose, you are likely to eat more in the future. In the short term, fructose doesn't suppress a hormone called ghrelin. A drop in ghrelin is what makes you feel full while you're eating. In addition, fructose is converted into fat up to fifteen times more readily than glucose is. In animal studies, fructose has been shown to contribute to all the symptoms of Syndrome X.

The data in humans is less clear possibly because research has focused on the effects of eating sugar, or sucrose. But we mustn't forget that sugar is made up of 50 per cent fructose and so has the same effect as consuming pure fructose in soft drinks.

Nutrition is still an inexact science. But there is increasing evidence to suggest that sugar contributes to the rise in obesity.

Firstly, it impairs appetite so that we eat more after we have consumed sugar. Secondly, sugar, in the form of soft drinks, is not 'noticed' by the body, which means we consume extra calories without being aware of them. Thirdly, eating sugar prevents fat from being oxidised, so that, although glucose in sugar is rarely converted to fat under normal circumstances, other food that we eat, particularly fat, is and fat already stored in our adipose tissue is not used. Fourthly, half of sugar is composed of fructose, which is implicated in diseases like those associated with Syndrome X and is readily converted into fat. Finally, sugar – the glucose part – can be directly turned into fat if we eat large amounts of it.

The ex-directors of Tate & Lyle I interviewed said that eating too much of anything can make you fat and that even water is bad for you if taken in excessive amounts. To some extent they are right – the old adage 'calories in, calories out' holds true. We all burn up a certain amount of calories a day but, if we consume more than we use up, we'll store the excess as fat. After all, eating just five teaspoons of sugar a day beyond what you would normally burn up, the same amount found in a small glass of Coca-Cola, half a Mars bar or a Kit Kat, will result in a weight gain of five kilos or ten pounds in a year. The problem is that sugar-rich foods are tempting, easy to eat and high in calories. Two sweet biscuits contain more calories than half a kilo of carrots, and which would most of us prefer to eat?

To some extent, Yudkin is vindicated, though not in perhaps the way he originally formulated his thesis. Sugar does contribute to obesity; fatty acids produced by sugar can lead to increased blood pressure, coronary disease and insulin resistance. Yudkin also mentioned skin problems caused by sugars: the Sydney group has found that high-GI foods are associated with heightened levels of acne. Yudkin said that sugar caused gallstones: a study published recently in the *American Journal of Clinical Nutrition* showed that a lack of exercise combined with a high intake of sugar and saturated fat led to gallstones.

But not even Yudkin could have predicted some of sugar's other effects. One burgeoning area of research examines the formation of free radicals. During the normal course of metabolism, the food we eat is broken down into free radicals. These are single oxygen atoms, or hydroxyl groups (like the oxygen-hydrogen pairs on a sucrose molecule), which are highly reactive; when they interact with other molecules they can set off chain reactions, each one triggering the release of yet more free radicals. Some of these reactions are with the chemicals, enzymes and genetic material in the very cells of our body; they are, therefore, highly damaging to the body and are thought to be responsible for causing some types of cancer and contributing towards ageing. Free radicals can also increase inflammation and the accumulation of plaque in arteries that leads to cardiovascular disease. Antioxidants neutralise their effects – we all have natural antioxidants, but we need to obtain more from our diet. They're found in vitamins A, C and E, for example.

In a study published by the *Journal of Clinical Endocrinology and Metabolism*, volunteers were either given a drink of non-sweetened water or water with 75g of glucose added to it – about the same amount found in two cans of cola. The research group discovered that the people who consumed glucose had more than twice as much free-radical generation as those who had sugar-free water. Even more worrying, vitamin E activity – the very antioxidant that could suppress the damage – was depressed for more than three hours in those who had consumed the glucose.

On top of sugar's effects on the body, however, is the question of what it does to the brain. After Patrick Holford's conversion to big salads and soya sausages by the Wrights, he started to investigate the impact of nutrition in his own field: psychology. He discovered that eating the right foods could have a greater impact on mental-health problems, such as schizophrenia, than drugs and psychotherapy combined. Holford heard of a brilliant American academic, Dr Carl Pfeiffer, founder of Princeton's Brain Bio Center, who had studied brain chemistry. Pfeiffer had a massive heart attack when he was fifty. He was given ten years, at the most, to live, *if* he had a pacemaker fitted. He decided not to and to pursue a diet of optimal nutrition – for the next thirty years. Pfeiffer then passed on his knowledge to his schizophrenic patients and achieved a remission rate of 80 per cent. He claims that blood-sugar imbalance is one of the five main underlying factors in schizophrenia. Holford visited him in America and now, after years of working as a nutritionist, he says, 'I would guess that well over fifty per cent of people with mental-health problems, from depression to schizophrenia, have blood-sugar problems as a major underlying cause.'

Moreover, sugar can affect brain functioning in other ways. In a study carried out at a Scottish mental institution, half the patients were allowed unlimited access to sugar and the other half were not. Within two years the mental capacity of the group that had eaten sugar had significantly deteriorated in comparison to that of their peers. In 1983 researchers at Massachusetts Institute of Technology (MIT) discovered that high sugar consumption is also linked to IQ. They found that children who had a high intake of sugars and refined carbohydrates had an average IQ score 25 points below those who consumed a low amount of sugars. (However, in this case, correlation does not necessarily imply causation. An alternative interpretation is that sugar intake could be related to a generally poor diet and that this may be a reflection of a low-income or deprived household that would not be conducive to learning.)

Sugar has one more major effect on the brain: it can lead to addiction. Holford explains:

> If you eat a lot of sugar, your brain releases the chemicals dopamine and serotonin, which make you feel good. Sugar gives you a high – you get a buzz from eating sugar. But if you continue to eat large amounts, you become less sensitive to dopamine, in exactly the same way as you do when you have any addictive substance. It's called 'downregulation'; as a consequence you will have a sugar high, then a low and after that you'll crave more sugar to get the same effect. If you stop eating sugar, you start suffering withdrawal symptoms. This is the hallmark of an addictive substance and it's one of the big *bêtes noires* for the sugar industry.

So much for the theory – but is there any evidence that sugar is truly addictive? In 2002 researchers published work in the journal *Obesity Research* showing that sugar caused most of the same effects as addictive drugs. Professor Bart Hoebel, a psychologist from Princeton University, says that sugar triggers the production of the brain's natural opioids and that, 'We think that is a key to the addiction process. The brain is getting addicted to its own opioids as it would to morphine or heroin. Drugs give a bigger effect, but it is essentially the same process.'

Hoebel says that three characteristics have to be present for a substance to be classified as addictive: firstly, there is a pattern of increased intake and changes in brain chemistry; secondly, there are signs of withdrawal and changes in brain chemistry upon withdrawal; and, thirdly, there are signs of craving and relapse after withdrawal. Hoebel and his colleagues fed rats a nutritionally balanced diet as well as increasing amounts of sugar first thing in the morning. They then stopped giving them any sugar. The rats' teeth started chattering, a common sign of withdrawal, and they showed anxiety as well as a change in the balance of brain chemicals. His experiment, therefore, demonstrated the first two signs of addiction (though the increased intake was aided by the scientists themselves), leading him to suggest that the rats were sugar-dependent, but he is currently investigating whether they show craving and relapse too. As far as this work applies to humans, he says, 'It does change the way a person might look at it. It puts sugar in the realm of an addictive disorder, rather than a failure of willpower.'

Stronger comments come from Dr Candace Pert, a professor at the Department of Physiology and Biophysics at Georgetown University Medical Center in Washington, who studied the central role that endorphins play in addiction. She says, 'I consider sugar to be a drug, a highly purified plant product that can become addictive. Relying on an artificial form of glucose – sugar – to give us a quick pick-me-up is analogous to, if not as dangerous as, shooting heroin.'

It is little wonder then, that Holford calls sugar an 'antinutrient'. Sugar is not just devoid of goodness – it lacks any vitamins or minerals – but it also takes nutrients from the body in order to be digested. 'For every unit of sugar, you need vitamins B and C to turn it into energy; the more sugar you eat, the more you use up vitamins B and C,' says Holford.

None of the problems described above would be of overwhelming concern, however, if sugar consumption was falling. However, although consumption of packets of sugar has fallen, more and more sugar is currently being added to processed foods. Each year 2.3 million tons of white sugar is consumed in the UK; three-quarters of this is added to processed foods. In fact, 80 per cent of all refined sugar in this country ends up hidden in convenience foods and drinks. As Jeremy Smith, deputy editor of the *Ecologist*, says, 'We may be putting less sugar in our food and drinks, but that's only because someone has added more than enough for us already.' Biscuit manufacturing has risen by a fifth in the last thirty years and confectionery production has increased by 15 per cent in the same period. A survey carried out by the Food Commission in 2002 found that only two children's breakfast cereals contained less than 20 per cent sugar. Most brands were between 30 and 50 per cent sugar. Nearly all cough medicine contains sugar; Boots Children's Cough Relief was 66 per cent sugar, for example. Weight Watcher's low-fat salad dressing contained more than 10 per cent sugar, as did Safeway's curry-style savoury rice. Heinz Original Sandwich Spread was more than a fifth, and Bachelor's Cup-a-Soup was a whopping 37.5 per cent. Sweets and chocolate were, of course, even higher. Drinks also contained high levels of sugar: Coca-Cola had seven teaspoons per serving; Sunny Delight, ten; half a pint of beer, almost three. Because of the fat scare, many products are branded as 'low fat', allowing manufacturers to add sugar to alleged slimming products: the weight-loss drink Slimfast was 61.9 per cent sugar; McVitie's Go Ahead Fruit-Ins biscuits were 45.8 per cent sugar.

The problem, of course, is that we like sweet foods, but we are also aware that too much sugar is not good for us. This has resulted, according to Smith, in companies who pack our everyday foods with sugar not wanting us to know how much sugar there actually is in a product. Sugars are often listed by type (for instance, sucrose, glucose, fructose, maltose, dextrose, raw sugar, liquid sugar, demineralised sugar, corn syrup, high-fructose corn syrup, hydrolysed starch, hydrolysed sugar, syrup, and so on), leaving out a figure that relates to the total sugar content.

So, for example, Cadbury's Boost bar – 'charged with glucose' – includes glucose, dried glucose syrup and sugar as a component of the biscuit, as part of the milk chocolate and as glycerol. 'At no point is the consumer told quite how much sugar all these different types add up to,' says Smith. The company lists carbohydrates, not sugar, in the nutrition information. There are 59.6g of carbohydrates per 100g of Boost but the bar weighs only 60.5g. Even if you took a calculator to the sweet shop, you still wouldn't know how much of those carbohydrates were sugars. The average chocolate bar contains up to 70 per cent sugar – approximately 8 teaspoons or 240 calories worth of sugar.

Due to consumption of added sugars, the average British person consumes approximately 30 kilos per year. One in ten adults eats over sixty kilos (over nine stone) of sugar a year.

So who is to blame? Well, partly at least, we are. We buy the stuff, after all. But, even if you are an inveterate ingredient-reader, it is hard to tell exactly how much sugar is in a product or even why savoury rice or pasta sauce *needs* added sugar. Most of us have busy lives and cannot always cook using fresh ingredients. But this does not mean we want to consume vast amounts of sugar. In addition, portion sizes have vastly increased – the average muffin, for example, has swelled by 400 per cent in the past twenty years.

Are the sugar companies responsible? 'It's certainly not driven by the sugar industry,' retorts Saxon Tate. 'Apart from what you take and use in your kitchen, the sugar industry is not the end user or provider. The manufacturer [of the food] is the main provider. This is not to say that I would hide behind the manufacturer. The manufacturers are hopefully driven by what their customers want.' Saxon becomes quite heated. 'Don't buy those things with sugar in. If you don't want to have sugar in your diet, don't buy them. There's lots of stuff for you to buy.' Actually, there isn't. Unless you want to buy every ingredient fresh, it is almost impossible to find ready meals that are sugar-free in a super-market.

Why is this? If you were making fresh tomato soup, would almost half of your ingredients be refined sugar? The last time you cooked a risotto, did you add several tablespoons of sugar? If you were giving your child porridge in the morning, would you make sure half the bowl was filled with sugar? But food manufacturers, Saxon argues, are in a difficult position. They cannot reduce sugar because their customers like and have got used to sweetened food.

The situation is not quite that simple. In 1979 Professor Philip James, deputy director of the Medical Research Council's clinical nutrition department in Cambridge, was appointed by the government to chair a committee on national dietary guidelines. The committee recommended nothing radical – cutting consumption of saturated fat, salt and sugar. It linked sugar with dental caries, obesity and diabetes. However, James was criticised by the British Nutrition Foundation (BNF). The BNF's member companies include, among others, Tate & Lyle, British Sugar, Cadbury, Coca-Cola, Kellogg's, Masterfoods (who make Mars bars), McDonald's, Nestlé and Procter and Gamble. James felt that the criticism was unfair: 'Confuse the public, produce experts who disagree, try to dilute the message and indicate that there are extremists like me involved in public health,' says James.

In 2000 Aubrey Sheiham and other scientific experts were consulted on the formulation of the Eurodiet, a set of guidelines to help doctors and dentists advise their patients on healthy eating. One of the recommendations Sheiham put forward was that no more than 10 per cent of calories should come from sugar. In the end, however, what was published was that people should eat sugar no more than four times a day. Sheiham believed that this was too much, but felt that he and his colleagues had been pushed into this position. He says, 'The main thing is that the food industry and sugar industry want to create controversy. It's in their interests to contest anything the scientist puts out if it's against them.'

The Sugar Bureau drew up a report, in Sheiham's words, three times as long as his own to counteract it, although he felt that it did not deal directly with the issues he raised. According to Sheiham, many of the people at the conference who opposed his recommendations were scientists funded by the sugar industry.

At the Eurodiet conference, the ex-chief dentist of England stated that dental decay was no longer a problem in England. Sheiham's reply was to ask why, in that case, there are so many dentists. He adds, 'We have unacceptably high rates of decay and teeth are the most expensive part of the body to treat. We're constantly going to the dentist, but we don't have to visit the doctor with the same regularity.' Sheiham says that the chair of the conference, Professor Michael Gibney, from the Institute of European Food Studies, Trinity College, Dublin, Ireland, put pressure on Sheiham to alter the report. At the time Gibney had obtained 1.4 million Euros for research from Danone, Unilever, Nestlé, Craft, Coca-Cola and Masterfoods. Sheiham says, 'If I criticise his work – which was flawed – I criticise him and that's not good when he's the chair.'

Gibney responds, 'He's a big boy, for God's sake. Basically speaking the arguments he put forward I disagreed with.' Gibney's argument was that, if a person reduces the amount of fat they eat, they will increase their sugar intake. He felt that asking people to reduce sugar to 10 per cent of their total daily calories would mean encouraging them to eat more fat. He adds that he doesn't understand why Sheiham felt pressurised. Sheiham's opinion is that the way the sugar and food industry exert their influence is by funding most of the nutrition labs in the world, pointing out that this cannot but help make one biased: 'Whose food I eat is the song I sing.' He adds, 'I could raise a lot more funding if I were prepared to take money from the sugar industry.'

Sheiham had had a previous run-in with the big boys when he wrote a report entitled 'Sweet Nothings', discussing what he sees as the incorrect information companies disseminate about sugar to health professionals as well as ordinary people, from leaflets in hospitals to advertising on television. He was asked to write the report and was funded by the Health Education Council (HEC), but the Sugar Bureau, via an MP lobbying on their behalf, 'suggested' to the government that the report be taken out of the HEC's own library. Sheiham paid for fifty copies to be printed and circulated them to the press. The Sugar Bureau's current director, Dr Richard Cottrell, says that he has no idea whether Sheiham's version of events is right but questions his ability to write such a document given that he is a dentist, not a nutritionist. Referring to 'Sweet Nothings', he adds, 'It was incorrect then and it is incorrect now. The truth is that it is the frequency of sugar intake, not the total amount, that affects dental health. The claims that sugar dilutes minerals from the diet, and that it contributes to heart disease and obesity, have all been rejected by various committees. What is the relevance [of 'Sweet Nothings'] to anyone other than Aubrey [Sheiham]?'

Dr Jack Winkler was another to fall foul of the food companies. Winkler is a sociologist from New York who teaches nutrition policy at University College London and London Metropolitan University. In the early 1990s he became chair of Action and Information on Sugars (AIS), a group set up to dispel myths about sugar propagated by food companies. The first major campaign AIS took on was aimed at drinks for babies that contained sugars.

In the 1950s SmithKline Beecham (now GlaxoSmithKline and hereafter referred to as GSK) launched a drink called Baby Ribena. Shortly afterwards other manufacturers started producing baby fruit juices and sweetened herbal teas. As the sale of these products rose, dentists noticed something officially referred to as nursing-bottle caries, though dentists themselves often call it Ribena caries. When a baby sucks a bottle, the teat is pressed against the front teeth and the liquid is squeezed out just behind them, thus exposing them to sugar. Nursing-bottle caries, therefore, are seen when a child's front teeth, instead of their back teeth, decay first. In the worst cases, the teeth decay before they even erupt, and there's a brown line across the gum where the diseased roots are.

Although this is a recognised condition, no records were kept of the numbers of teeth destroyed in this way, but data are held of the number of general anaesthetics carried out to extract teeth. Winkler estimates that at the time there were at least 100,000 cases of nursing-bottle caries a year. As a result of AIS's campaign, sales of sweetened drinks for babies fell. In 1991 Winkler found out what had been happening behind the scenes. Parents had been complaining to GSK and had threatened to sue the company. Most cases were settled out of court with an attached no-publicity clause. But a case with ten parents did reach court. A reporter picked it up and it was all over the *Sunday People* the next day. Winkler helped parents form a legal action group to sue the company.

Sales of baby drinks fell by a third and companies attempted to reformulate their products. Sources within GSK told Winkler privately that they were developing a new drink that would be safe for babies. Although, in fifteen phone calls to the company over three years, Winkler expressed his support of the company's policy, he was not aware of the launch of Ribena Toothkind in 1999 until Channel Five called to ask his opinion on it. Looking at the ingredients, Winkler believed instantly that this product was not going to be in the least kind to teeth. Blackcurrants are tart, despite the natural sugar they contain; in order to make a blackcurrant drink palatable, it has to have some form of sweetness added – and the drink did indeed include artificial sweeteners. It is illegal to sell sweeteners to children under four but the marketing and advertising aimed the drink at children and, Winkler believed, at babies. Winkler realised it would cause tooth decay, even though the drink had been accredited by the British Dental Association (BDA) and GSK explicitly claimed that the drink did not 'encourage tooth decay'.

Winkler phoned a friend of his in Switzerland who runs the world's leading laboratory that accredits tooth-friendly products. He had not heard of Ribena Toothkind either. The two men decided to test Toothkind for themselves. The drink failed the tooth-friendly test, quickly producing more than the recommended amount of acid in the mouth. Winkler used this evidence to complain to the Advertising Standards Authority about GSK's claim that the drink did not 'encourage tooth decay'. The BBC, working with Winkler, sent another sample of Toothkind to the Zurich lab, only to discover that the person who carried out the procedure had already conducted the same test for GSK when Toothkind was under development. The Ribena drink had failed that time, too.

GSK insisted on four rounds of evidence, two outside-expert assessments and two appeals, stretching the case out over the course of two and a half years. In the end, GSK took the case to the High Court where the judge found GSK's claim misleading and thus illegal. The company was not fined but sales of Ribena Toothkind fell by 15 per cent, resulting in a loss of revenue to GSK of £3 million. The BDA did not remove their accreditation, but the claim about tooth decay has now been eliminated from all products and advertising. It was the AIS and Jack Winkler's greatest success.

Although many food manufacturers are now printing health advice on their packets or websites on, for example, the dangers of excess sugar or the benefits of eating small amount six times a day with healthy snacks, food companies can fudge information – and not all snacks promoted within the food industry could be described as healthy.

Millions are spent on advertising each year – more than £200 million in the UK by leading food manufacturers. Much of it is aimed at children. Naturally, children like sweet foods; the more they eat, the higher the profit for food companies, but also the more likely these companies are to have a customer for life. Between 1991 and 1996, expenditure on children's food advertising doubled, closely followed by sweetened breakfast cereals. One survey found that six brands – Coco Pops, Sugar Puffs, Milky Way, Wispa, Maltesers and Mars bars – comprised a quarter of all food advertised during twenty hours of children's television. Another study has calculated that food advertising accounts for almost half of all the advertising shown during children's television in thirteen countries (Australia, Austria, Belgium, Denmark, Finland, France, Germany, Greece, the Netherlands, Norway, Sweden, the UK and the US). Six of these countries allowed sponsorship of programmes by food companies, usually promoting confectionery, soft drinks and sweetened breakfast cereals.

One infamous campaign to produce educational material for schoolchildren, which was banned, was created during the early 1980s. One video was targeted at helping children understand what foods caused tooth decay. The answer given was 'bread, rice, crisps and cucumbers'. No mention of sugar. And cucumbers? The video's presenter says, 'Did you know that a cucumber is 80 per cent sugar?' (This figure was taken from the *dry* weight of cucumbers.) If cucumbers were 80 per cent sugar, we'd be refining them instead of sugar cane.

In 1983 Saatchi and Saatchi won a silver award for the most out-standing public service campaign for an ad commissioned by the HEC, which encouraged people to reduce fat, increase consumption of dietary fibre and cut back on sugar. The campaign was launched in part because, in 1982, 374,000 general anaesthetics had been administered to children under the age of fifteen for tooth extractions. The HEC claimed most of these could have been prevented by better dietary habits. However, in spite of the award, the ad was never run.

Ironically, less than a year later British Sugar hired Saatchi and Saatchi to create 'the biggest ever advertising campaign by a sugar manufacturer ... based on the theme that sugar is fundamental to taste enjoyment and is an essential ingredient in a healthy, active life.' A million pounds was devoted to it and the target was that 80 per cent of housewives would see the ad at least ten times, thereby increasing sales of speciality sugars, such as brown sugar, by a third.

Perhaps one of the most insidious ways people are initially encour-aged to eat sugar is by giving it to them when they are babies. We know advertising is mainly aimed at children, but children can be per-suaded to adopt a sweet tooth even before they can talk. Between a fifth and a third of many baby rusks is sugar and baby foods routinely have between one and three teaspoons per serving. John Yudkin pointed out that overfeeding babies will not only make them fat, but will also encourage the growing child to lay down even more fat cells. Once a fat cell has been formed, it can be filled or emptied of fat, but it will never disappear. He concluded:

As an adult, you will be one of those miserable fat people who can diet only with great difficulty, and by going hungry, and get their weight right down to what it should be – only to regain it with demoralising speed. Fat babies not only grow into fat adults; they grow into fat adults with real slimming problems.

So should the government legislate to reduce the amount of sugar in food or should it be the food manufacturers' responsibility? Are consumers to blame for doing too little? Saxon Tate says, 'I find it totally unacceptable that I should be lumbered with the responsibility for telling people what they should or should not eat. I'm willing to accept legislation and I'll happily accept changing taste. I do not feel I have a moral duty to tell people what they should or should not do.'

He would not, however, encourage legislation: 'It's not the role of government, as far as I'm concerned, to interfere in what individuals eat or don't eat. I am so horizontally opposed to that sort of interference with my freedom.' He does agree that people are consuming too much sugar, but believes that the change must come from parents educating their children to eat more healthily. In contrast, fellow retired director Colin Lyle says, 'If I was a food company, and sugar was seriously implicated as a health hazard, I would feel it morally and commercially incumbent on me to reduce the sugar content.'

Toni Steer summarises the situation by saying:

I think the food industry partly needs to take responsibility, but it's trapped between the devil and the deep blue sea, because, if it reduces sugar content, customers will say their food tastes horrible and they'll lose money. They should gradually reduce sugar and label products more clearly. Individuals should also take more responsibility and make an attempt to reduce intake ... The food manufacturers will have to make some voluntary changes, otherwise they will be faced with legislation.

If we should reduce our sugar intake, whether by our own free will or because the government or food manufacturers have responded to public demand and health advice, the difficulty would be in deciding how much a healthy intake of sugar would be. People such as Patrick Holford would say that we don't need any refined sugar at all, although he says he is not 'rigidly obsessive' and would eat a slice of cake at a tea party. We can certainly obtain enough sugars and energy from eating food that does not contain any refined sugar at all, but most of us would not like to live under such a restrictive regime.

In the 1990s Aubrey Sheiham published a scientific paper in which he examined dietary guidelines published throughout the world during a thirty-year period between 1961 and 1991. He looked at 115 reports from 36 countries, both developing and developed. The majority (84.5 per cent) proposed that sugars and, in particular, sucrose should be limited. Most considered that the maximum amount of sugar we should eat should be not more than 10 per cent of our total calorie intake. So, for example, if the average man needs 2,500 calories per day, he should not eat more than 250 calories of sugar (approximately 8 teaspoons), and, if the average woman needs

2,000 calories a day, no more than 200 calories (6½ teaspoons) should come from sugar. A bowl of breakfast cereal (with no added sugar), a portion of jam, a bowl of soup and a glass of wine would be the maximum amount for a woman – unless she wanted to blow it all on two-thirds of a Mars bar or a couple of fizzy drinks. The average person in developed countries far exceeds these guidelines: in the UK and the US we eat at least twice as much – the average American adult consumes 20 teaspoons, while male teenagers get through 34 teaspoons a day. Sheiham's paper was published in 1994 and remained quietly unnoticed, published, as it was, in Portuguese.

What has created more of a furore was exactly the same advice given by the World Health Organisation (WHO). In spring 2003 the WHO and the Food and Agriculture Organisation (FAO) published a report entitled 'Global Strategy on Diet, Nutrition and Physical Activity'. It was a response to the widespread decline of physical activity in most societies combined with rising fat intakes and consumption of energy-dense foods, which has led to an estimated 300 million obese and 750 million overweight people worldwide and, as a direct consequence, increasing levels of cardiovascular disease, cancer, diabetes, obesity, osteoporosis and dental disease. In 2001 these diseases contributed to 59 per cent of the 56.6 million deaths worldwide and 46 per cent of the global burden of disease. Within a decade, says the WHO, there will be a global epidemic if serious health action is not urgently taken. According to the *Observer* newspaper's health correspondents, Jo Revill and Paul Harris, the WHO report is 'the world's only realistic attempt so far to help poorer nations avert the looming public health disaster threatened by the growing burden of heart disease and diabetes'.

Developing countries, such as Brazil and India, are showing ever increasing rates of obesity. Even more than the developed world, populations in developing countries are at risk from junk food because they carry 'thrifty genes', which predispose a people to put on weight quickly if they have lived through centuries of near starvation and are then suddenly, within a few generations, faced with a plentiful supply of food. It is estimated that by 2025, if nothing is done, India will have one of the highest rates of Type II diabetes.

A panel of thirty independent experts working with thirty of their peers assessed the best scientific evidence available to compile the WHO/FAO report. They recommended that, of our total energy intake, fat should be reduced to 15–30 per cent, saturated fat no more than 10 per cent, sugar less than 10 per cent and protein 10–15 per cent; we ought to eat less than 5g of salt and more than 400g of fruit and vegetables, and take an hour's moderate exercise every day.

In early 2004 the report was debated in Geneva, but so far America is opposing its adoption. A review of the report by the American Department of Health and Human Services claimed that it did not meet their standards for scientific evaluation, nor did it separate science from policy-making. In its words:

> The WHO/FAO Report tends to mix policy and science conclusions, thus undermining this important principle designed to protect the integrity and credibility of the scientific review process . . . There is also an unsubstantiated focus on 'good' and 'bad' foods, and a conclusion that specific foods are linked to diseases and obesity (e.g., energy-dense foods, high/added-sugar foods, and drinks, meats, certain types of fats and oils, and higher fat dairy products). The USG [United States Government] favors dietary guidance . . . that promotes the view that all foods can be part of a healthy and balanced diet, and supports personal responsibility to choose a diet.

One of America's main objections is with the amount of sugar we should be eating: the US food industry have stated that up to a quarter of our daily energy needs can be safely supplied by refined sugars. Naturally, some of the people who support this assertion and oppose WHO work for the sugar industry: the sugar barons of Florida, the state in which most US sugar production is based. Food manufacturers worldwide, including some of the sugar industry, have also been directly lobbying the WHO, threatening to withdraw financial support.

Unsurprisingly, in America, President George W. Bush's initial reaction has been to say that exercise and eating sensibly is a matter for the individual, not the state. He argues that there is little evidence to show that drinking sugary drinks or eating sugar causes obesity. George Bush senior's godson, Bill Steiger, is the representative of the US Secretary of Health and Human Services, Tommy G. Thompson. He says that there is no evidence to indicate junk food makes children overweight. Yet half of all American children are obese and they didn't get that way by eating spinach. Professor Kaare Norum, from Oslo University, Norway, who headed the group of scientists advising the WHO on its strategy, counters that if the US does not support the strategy there will be an 'obesity bomb'.

Sugar, as Colin Lyle says, is all about politics. But those politics are now about to affect the health of the world's population.

10. SWEET SERENDIPITY

Everybody's singing the Sugar Blues
I'm so unhappy, I feel so bad
I could lay me down and die.
You can say what you choose
but I'm all confused.
I've got the sweet, sweet Sugar Blues
More Sugar!!
I've got the sweet, sweet Sugar Blues.
  'Sugar Blues', Lyrics by Lucy Fletcher,
    Music by Clarence Williams, 1923

Dr Mike Lindley, Director of Lintech, a company based at Reading Scientific Services Limited, who used to work for Tate & Lyle's research laboratory, says, 'The paradox is that we enjoy [sugar], it's an innate pleasure but we feel very guilty about it and it may not be nutritionally the best macronutrient we could be consuming … As a consequence that has driven loads of research programmes in lots of companies to find ways of delivering the physical and sensory characteristics of sugar without delivering all the nutritionally bad things – dental caries and calories. That was a large part of Tate & Lyle's philosophy.'

It is interesting that a company that denies any harmful effects of sugar should spend decades and millions of pounds attempting to create a product that mimics sugar but is not 'nutritionally bad' for you. The product that Tate & Lyle created has already been launched in the United States and has just been released in the UK. It is, in many respects, a dream product for Tate & Lyle.

The story began back in the 1960s with Saxon Tate. At the time Saxon was running Tate & Lyle's Canadian subsidiary, Redpath. He was the first Englishman to be elected to the president's advisory committee of the American Management Association. He says, 'I sat and listened to the bosses at General Foods, Coca-Cola, whoever, absolutely slamming the sugar industries' attitude towards artificial sweeteners. In those days, beverages were only sweetened with sucrose, and the association that represents the sugar industry in the United States had taken a completely negative attitude to the production of high-powered sweeteners, such as Nutrasweet.'

In 1967 Saxon wrote to Antony Hugill, director of research at Tate & Lyle and author of *Sugar and All That*. 'I said, "Look, you've just got to stop this. We've either got to buy in artificial sweeteners or we develop our own. What I'd like is one that is organic and based on the sucrose molecule."' This is one of those cases where you should be careful what you wish for as it might just come true.

Shortly after Saxon's letter, Colin Lyle inherited the responsibility for running the research lab. Almost immediately, the head of the laboratory died and it was suggested to Lyle that the person he should hire as a replacement was a Greek-American called Chuck Vlitos.

Vlitos, originally from Anatolia, came to America in the 1920s to study at Columbia University. He has a PhD in plant physiology and biochemistry and was recruited to run Tate & Lyle's research laboratory in Trinidad, which he did with considerable success, turning it into a world-famous institution. While in Trinidad he developed a spray from a plant hormone that could increase the amount of sugar in cane and was responsible for developing a variety of sugar cane, 41227, that he still remembers fondly today.

'I don't think I looked at anyone else. You choose an insider where you can,' says Colin. Bringing Vlitos over to the UK was the start of a lifelong friendship between him and Colin, which has outlasted the creation and dismemberment of Tate & Lyle's research laboratory, the early 'retirement' of both men from the company, and led to the formation of their own rival sweetener company as they step briskly into their 80s.

'Chuck,' says Lyle, 'was absolute tops as a research director.' Technically Vlitos was trained as a plant physiologist, but he also knew a lot about chemistry, including synthetic carbohydrate chemistry. Completely self-confident, Lyle says that he was always keen to give others credit. 'He never needed to have any glory taken from you to him. And he's also a very nice man, full of energy and with a great imagination: easy to talk to, open-minded, terrific.' Vlitos was also very good at self-promotion and persuaded the company to spend vast sums on research and development.

Vlitos, with characteristic energy, oversaw the move from Kent to

Reading University's campus, setting up what was the first commercial laboratory at a university in the UK. 'It was a very exciting time, a very creative place,' says Vlitos. 'We were pioneers – students came and worked with us; we ran co-operative projects with other universities. Our work was published and circulated worldwide; at one time we had thirty-five visiting scientists representing eighteen different nations.' The science was innovative: the first sweetener derived from a protein was isolated from the berry of the West African serendipity tree; research was conducted into the Jerusalem artichoke, a knobbly, potato-like root with incredible flatulence-inducing properties, to see if its flesh held a low-calorie sweetener. Under Vlitos, xanthan gum was developed, a chemical with such versatility that it is used as a lubricant when drilling oil wells as well as to thicken ice cream and prevent yoghurt turning watery. His team created a new detergent from a sucrose ester, which is edible, so does not harm the environment, doesn't sting the eyes and doesn't leave a nasty ring round the bath.

At the end of the 1960s and the start of the 1970s there was an energy crisis: it looked as if supplies of crude oil were going to run out. Vlitos felt that the solution might lie with sugar – instead of just eating the stuff, we should use it as a fuel, or create new materials that could be used as an energy source. After all, Tate & Lyle had long marketed it as a source of 'pure energy' for fuelling the human body. 'What we wanted was a new raw material that could be produced biologically, year after year, without harming the environment,' says Vlitos.

He put between sixty and seventy people on the project, some of whom were working at Queen Elizabeth College, London. This group was run by Professor Leslie Hough and his brilliant Pakistani post-doctoral student, Dr Rhiaz Khan. Khan made several hundred chemicals by altering the structure of the sucrose molecule itself. 'We had no idea what applications they would have,' says Vlitos, 'I thought we should test them for fungicides, pesticides, furniture polish, whatever.'

The story goes, and it may or may not be true, that in 1972 Khan asked one of his researchers, Dr Shashikant Phadnis, whether he'd tested a particular molecule and that Phadnis misheard and thought he said, 'Have you tasted it?' He replied that he had not, but would do so right away. On tasting a tiny amount of this manipulated sugar molecule, Phadnis realised it was sweet and phoned Khan back to tell him.

In fact, it was more than sweet, it was explosively so – the team had created an imitation of sugar that was 600 times sweeter than the real thing.

In 1976 the group published their work in the prestigious scientific journal *Nature*. Tate & Lyle may not quite have realised the full potential of what they had on their hands, but someone else did. Within days, Vlitos had a phone call from the United States. It was a Dr Wallis Steinberg from the megacorporation Johnson and Johnson, asking to meet him at New Jersey airport. 'He said he wanted to talk about buying the rights to this new sweet compound of ours.' Vlitos laughs at the memory. 'At the time we didn't have very much of it – we had enough to put on the end of a pin.'

Vlitos did see Steinberg, but suggested at the next Tate & Lyle board meeting that they shouldn't just give away their new compound. Colin Lyle agreed, and also concurred with Vlitos that Tate & Lyle did not have any experience in food safety or how to go about testing it. Tate & Lyle had already had its fingers burned by buying up businesses it knew nothing about; Colin had always been opposed to this strategy, even to the extent of believing that the merger with Tate back in 1921 had been a mistake. So Colin brokered a joint-venture agreement with Johnson and Johnson whereby the latter would conduct toxicology tests and Tate & Lyle would produce the new chemical.

The first version of this new chemical was quite different from its final formulation, but essentially what the scientists had done was to substitute three of the oxygen-hydrogen groups on the fructose and the glucose molecule with chlorine atoms. It was, in essence, a chlorinated sugar molecule. Its proper name is 1,6-dichloro-1,6-dideoxy-[beta]-D-fructofuranosyl-4-chloro-4-deoxy-[alpha]-D-galactopyranoside. They called it sucralose.

It was, to Tate & Lyle, a miracle product – it is made from sugar, so it sounds 'natural' to the consumer; it has the crystalline appearance of sugar; it is heat stable (so it can be used for cooking and baking); it has a long shelf life; it dissolves in water; it is 600 times sweeter than sugar; it did not look as if it would cause dental caries and, above all, it has no calories. It simply delivers its devastatingly sweet taste and passes straight through the human body. All the fun and taste and usefulness of sugar, but without the guilt. What was more, it looked as if it would beat the competition hands down.

In the 1960s, sucrose's main rival was cyclamate. It was banned in

1970 as it allegedly caused bladder cancer, but the US Food and Drug Administration (FDA) is considering whether to grant approval again.

Sucralose's other competitor was saccharin: developed in 1879, it is 300 times sweeter than sugar. According to the FDA, it was considered 'GRAS' – 'generally recognised as safe' – but in the 1970s the agency decided to retest hundreds of products labelled GRAS. In the year sucralose was discovered, the FDA found that saccharin caused bladder cancer in rats. The tests were repeated in 1973 as there was some concern that impurities, rather than saccharin itself, was responsible. However, in 1977 Canadian studies showed conclusively that saccharin was the cause. Saccharin is still added to products and is available as a low-calorie tabletop sweetener but it is required to carry the label: 'Use of this product may be hazardous to your health. This product contains saccharin, which has been determined to cause cancer in laboratory animals.' This label implies that there is no proof that, because saccharin causes cancer to rats, it will have the same effect in a person. However, a wide-ranging study carried out by the FDA in the late 1970s showed that anyone who takes large amounts – eating it six or seven times a day – could run an increased risk of contracting cancer.

In 1981 a new sweetener was developed that was 180 times sweeter than sugar – aspartame, sold under the brand name Nutrasweet. Aspartame has one big disadvantage compared to sucralose and saccharin – it breaks down when heated. Aspartame has FDA approval, no warning like saccharin and is considered by the agency to be 'one of the most thoroughly tested and studied food additives the agency has ever approved'. It is considered 'safe for the general population'. Nevertheless, there have been health scares associated with it and allegations have been made that it causes systemic lupus, multiple sclerosis, vision problems, headaches, fatigue, and even Alzheimer's disease and Gulf War syndrome.

According to the FDA, there is no way aspartame can be linked with Gulf War syndrome, multiple sclerosis or systemic lupus. However, when you eat aspartame, it turns into methanol, formaldehyde and formate, substances that are toxic at high doses. There are also two amino acids (proteins) in aspartame – phenylalanine and aspartic acid – that can cause brain damage when eaten in quantity. The FDA says that this will only be a problem for people who have the rare hereditary disease phenylketonuria, which affects 1 in 16,000 people. A rather less rare group that are also sensitive to phenylalanine are pregnant women with high levels of it in their blood. As a result, the FDA requires all products containing aspartame to have pheny-lalanine on the label. As for aspartic acid, which also has the potential to cause brain damage at very high doses, the FDA claims that it is unlikely that any consumer would eat or drink enough aspartame to sustain brain damage: FDA figures show that most aspartame users only consume about 4–7 per cent of the acceptable daily intake the agency recommends for the sweetener. Aspartame is currently part of an ongoing study by the National Cancer Institute in the States that is looking at diet and the onset of adult brain cancer.

Tate & Lyle had high hopes that sucralose would prove better than its competitors as far as safety was concerned. After all, it had all the hallmarks of a perfect sweetener. Aspartame is actually calorific but, because it is so sweet, only tiny amounts need to be eaten. Sucralose has no calories at all. Dr Mike Lindley says that its only flaw is that the sweetness of sucralose is perceived slightly later than sugar – just microseconds – and it remains sweet-tasting slightly longer in the mouth than sucrose. 'If you ate sucrose and sucralose side by side you'd notice it. It's very subtle, but it's a real difference,' he adds.

By the mid-1970s Tate & Lyle had hammered out their deal with Johnson and Johnson: Tate & Lyle would supply the sucralose and Johnson and Johnson would be in charge of clinical trials. Safety tests on the new sweetener could now begin. 'Johnson and Johnson were superbly run, a wonderful company with a great reputation and con-siderable experience at dealing with the FDA,' says Colin. 'I thought that this should help us enormously and I can remember in about 1976 saying we should forecast getting FDA approval in two or three years.' In fact, it took more than twenty years. Saxon says:

> I think any other company would have dropped it a long, long time ago. But Tate & Lyle is curious in that way. Once you start on something, unless you are convinced you've got it wrong, you stick to it. The question you have to ask yourself, and I'm not sure we've done this sufficiently, was will it prove to be in the shareholders' interest? I think the jury's out on that one.

FDA approval was not granted until 1 April 1998. Part of the problem was the use of chlorine – a chemical that is normally toxic when eaten. 'The FDA know they'll be criticised if they approve a drug or a food that subsequently kills people. They don't know they'll be praised if they approve a product that does benefit people. So you play safe, don't you?' says Colin. But he adds, 'My feeling there is that all the elaborate FDA testing cannot ensure one hundred per cent safety with everybody in the world.'

The point of toxicology testing is counter-intuitive. The idea is to find a toxic dosage, not a safe dosage. Once a toxic dosage has been determined, the safe dosage is considered to be a hundred times less than the amount that begins to cause a toxic reaction. But in order to work out what is toxic, animals have to be fed massive doses of the chemical until it has an effect. This is the origin of the infamous LD50 test – a product is fed to animals until a lethal dose is determined when half of them die. More than 113 experimental studies were carried out in the UK and the US. Over a period of twenty years sucralose was tested on mice, rabbits, rats, monkeys and dogs.

At the end of it all, however, sucralose was given a clean bill of health and FDA approval. The accepted daily dose is fifteen milligrams per kilo of body weight – for the average woman that would be just under a gram a day. Sucralose is on sale in over fifty countries, including Britain, where it is marketed as Splenda by McNeil Consumer Nutritionals Ltd, a Johnson and Johnson company. In the US it already commands a fifth of the total spent on sweeteners after only 18 months; in the UK the market is worth £52 million and McNeil have spent an enormous amount on marketing to launch the product, including television advertising that will probably appeal to Britain's 14.4 million dieters and 2.4 million diabetics.

Tate & Lyle have developed another type of sweetener – lactisole, a compound that *inhibits* the sweetness of sugar. It works by blocking the taste receptor in the tongue so that it is possible to eat sugar or other sweeteners without tasting them. Why, you might ask, would you want to do that? Mike Lindley explains:

Sugar is a very useful ingredient in that it develops a lot of characteristics which are of value, like bulk; it's a preservative; it depresses freezing points and drives water out of food and it's cheap, but, along with all those physical characteristics, comes sweetness. So you could, in theory, make a savoury version of all the products today which are sweet: you could have a savoury Mars bar, create avocado ice cream or a mouse-flavoured sugar mouse.

Lactisole, made from milk sugar, is incredible. Eat a pinch with a teaspoon of sugar and the sugar is as tasteless as sand. The catch, of course, is that you're still eating sugar but may not even be aware of it. It is also possible that you may be consuming more calories, depending on the food product it is in, than you think you are. The consumer is being fooled. 'Correct,' says Lindley. 'It is a concern that you could be deluding consumers.' Lactisole, marketed by Domino Foods, is now on sale in the US.

Lindley has given a number of reasons why sugar is such a popular food ingredient for manufacturers, whether or not we can taste it, but one of his main reasons for explaining why there is such a high amount of sugar in foods is simply because it is cheap. He says:

> You can buy jam that contains high concentrations of fruit and low concentrations of sugar, or you can buy jam that contains low concentrations of fruit and high concentrations of added sugar. The cheaper one has a high concentration of sugar and a low concentration of fruit and not everyone can afford the one with more fruit. It would be nice if we could all afford high-fruit, low-sugar jam, but sadly we can't.

It's a nice theory, except for the fact that, although sugar may be cheap in the supermarket, it has hidden and long-term costs – its true price is ruinously expensive.

11. BIG SUGAR AND THE RIVER OF GRASS

As far as the land is concerned, we are living on our toes. Anything could happen. We have got complete insecurity. The worst thing I could think of is to have to move from here. That would be worse than death.

.Mandelkosi Ndlovu, Sugar farmer from South Africa
in an interview carried out by CAFOD representative

Camille is 82 years old, but, unlike most people in the developed world who are his age, he has not retired. 'Look where I'm living. Look at the house I have – and I work every single day.' Camille is a sugar-cane cutter in the Dominican Republic. He lives in a wooden shack with no running water, electricity, toilet or anywhere to cook other than an open fire outside. He has no pension, is paid a pittance and does not even own his house. When he stops working, he will have to leave. According to organisations like the Catholic Institute for International Relations (CIIR), Europe's sugar policy established thirty years ago has contributed to Camille's poverty and that of thousands like him.

The 1975 Lomé Convention, and the continued subsidisation of sugar beet, directly affected Tate & Lyle. What is perhaps less well known is how this affects consumers of sugar, as well as developing countries. To understand why sugar is cheap in price – and therefore is in most of our food – but costly in terms of lives, we need to take a step back. After the Second World War, food shortages in Europe encouraged the member states of the European Economic Community (EEC, now the European Union – EU) to become more self-sufficient. This led to the creation of the Common Agricultural Policy (CAP) in 1958; Europe's sugar policy is part of the CAP and is designed to support sugar-beet farmers.

Once a year, a minister from each of the European member countries agrees a guaranteed minimum price for a set amount, or a 'quota', of sugar. This minimum price is normally three to four times higher than the world price and the quota itself is large – about 25 per cent more than Europe consumes. Because of this guaranteed subsidy, sugar-beet farmers are encouraged to produce excess sugar, which is then sold on the world market. World prices are inevitably lower than the cost of producing sugar in Europe, so European sugar is normally sold at 50–65 per cent of the prices guaranteed by the CAP. This practice is known as 'dumping'. In this case, the European Union pays the farmers the shortfall. The cost – just under a billion pounds – is ultimately paid by European consumers and taxpayers, and comes to around £42 per family.

France and Germany produce the most sugar and, therefore, dump the most but, relative to the amount of sugar they produce, Britain and France are the worst culprits, dumping a quarter of their sugar surpluses on the world market. The effect is that the EU depresses world prices, often below the price of sugar from the lowest-cost producers, undercutting countries such as Malawi, Mozambique and Zambia.

European sugar-beet farmers are further protected from world competition: if the price in world markets is below EU prices, imports are taxed to ensure they cannot undercut European prices. If the EU market price for sugar falls below the guaranteed minimum, then farmers can sell directly to EU agencies, who have to buy their sugar and will usually keep it in storage. Known as intervention buying, the aim is to raise the market price back to the agreed minimum, at which point it can be sold. If this doesn't happen, these mountains of sugar have to be sold at a loss or may even be disposed of.

Despite the fact that sugar beet is grown expensively and inefficiently – it costs well over 50 per cent more to produce sugar from beet than from cane – the result is that, instead of Europe importing sugar from countries that can grow sugar cane, Europe has become the world's second-largest sugar exporter (after Brazil), exporting 40 per cent of the world's white sugar. It costs Europe £364 to produce a ton of white sugar, compared to £154 from more competitive countries, such as Brazil and Zambia. Europe's sugar could not be grown if beet farmers were not heavily subsidised; in the UK beet-sugar farmers each receive an estimated £60,000 in subsidies each year.

In 1975 the Lomé Convention was also responsible for an arrangement with Europe's former colonies, who were allowed to export a set amount of raw sugar – 1.3 million tons – to the EU, and for this they were, and still are, paid the guaranteed minimum EU price for beet sugar, minus what it costs the UK to refine it for them. Seventeen countries make up this Sugar Protocol, divided between Caribbean countries, such as Barbados, and African countries, such as Mauritius and Zimbabwe. They are known as the African, Caribbean and Pacific (ACP) countries. The Caribbean countries tend to export 35–75 per cent of their sugar, which represents 10–25 per cent of their foreign exchange earnings. The cost of sugar production is high in these countries – according to research carried out by the CIIR, none of these countries has lower production costs than the most efficient EU producers – as their facilities are old and inefficient. Interestingly, even though Tate & Lyle had to sell most of their investments abroad, such as Trinidad and Jamaica, many former colonies, including these two, have asked the company back.

African Sugar Protocol countries (apart from Malawi and Zimbabwe) rely on sugar exports to a lesser extent – about 10 per cent of sugar produced is exported as most is kept for domestic and regional markets; exports represent 2.5 per cent of foreign exchange earnings. These ACP countries employ around 275,000 people, though many more are involved in related activities, or are dependent on those working directly in the sugar industry. In contrast, European cane refiners employ 3,000 people.

At the time of the Lomé Convention, sugar prices were extremely high, which put the ACP in a strong position – Europe needed to import sugar, since it could not meet demand. But the heavy subsidies to the beet industry resulting in the EU becoming a net exporter of sugar and, thus, not needing sugar from the ACP countries mean that some, such as the European Court of Auditors, see the ACP as an 'unnecessary burden' on the CAP. Others view it is an outmoded way of donating aid rather than a trade agreement.

Countries who are not part of the ACP are virtually unable to export sugar to Europe at all since, firstly, the world market price has been depressed below the price it costs them to produce sugar themselves and, secondly, tariffs reaching 140 per cent are imposed on imports, making sugar one of the most protected of Europe's products.

So who benefits from Europe's sugar policy? Firstly, the beet-sugar farmers, who constitute only 4 per cent of all European farmers yet are the world's biggest recipients of sugar subsidies. Kate Raworth, author of Oxfam's 2003 Briefing Paper 'The Great EU Sugar Scam', says, 'The perverse result is that most sugar beet is grown where it is least suited: over half of Europe's quota is allocated to countries with below-average productivity rates.'

Secondly, the refiners benefit since the money in sugar is made from processing, not growing it. The refiners in the UK are Tate & Lyle (for imported sugar from the ACP) and British Sugar (for beet sugar) – British Sugar receives around £77 million in subsidies, and has an overall profit margin of more than 25 per cent. The British Sugar Corporation was created as a government monopoly in 1936. It was then privatised in 1981, becoming British Sugar, and was sold to Associated British Foods (ABF) ten years later. In 1990, British Sugar and Tate & Lyle planned to merge to create one firm, but this time the Monopolies and Mergers Commission put a stop to it.

The ACP countries benefit too: they receive a stable income and in recent years earned £270 million a year more than they would have made by selling sugar on the world market. But, if these countries (particularly those in the Caribbean sector) had not been so dependent on exporting sugar, they might have been able to diversify their economy. Mauritius, for example, has devoted more than 90 per cent of agricultural land to growing sugar and has the largest quota to the EU. In net terms Mauritius may be better off by growing what it can sell rather than by diversifying but, for the country to survive, diversification is necessary in the longer run. Here again, Europe has not been helpful. The Mauritian Government has made efforts to diversify and used money from selling sugar to finance a textile industry. Europe's response was to impose an export restraint on Mauritius because its clothing imports were seen as a threat.

Today in Mauritius sugar is still important, providing just over a fifth of revenue. But costs in the country have escalated, particularly labour, as mechanisation is difficult due to the hilly terrain and the number of small farms. More than three-quarters of the sugar estates are running at a loss. Plantation owners claim that they provide medical centres, education and free transport for children to get to school and sporting events. They allow trade union organisation and there is a pension fund. As a result, raw sugar production costs are higher here than in most other countries.

According to the CIIR, Mauritius, in spite of the government's care, is in a difficult situation.

Economic diversification has gone far enough to threaten the country's viability as a sugar producer, should there be a cut in the Sugar Protocol [ACP] price, by raising the costs of production; but diversification has not gone far enough to reduce the dependence of the Mauritian economy on sugar.

Those who do not benefit from the Sugar Protocol are countries that produce sugar efficiently and are not part of the ACP, such as Australia, Brazil and Thailand, and some of the least developed countries in the world, only four of which are included in the ACP. These are the most poverty-stricken places in the world: around one in twelve children die before the age of five; more than a third do not attend primary school; life expectancy is 63; half of all households lack access to safe water and sanitation; and malaria and HIV are widespread. Some of these countries, such as Malawi, Mozambique and Zambia, are the lowest-cost sugar producers in the world. Because of depressed export prices, many countries are suffering; Cuba, for instance, may have to close almost half of its sugar mills within the next few years.

Oxfam estimates that, without the CAP, Europe would import seven million tons of sugar, instead of exporting five million, and some of this sugar would come from the poorer countries and lower-cost producers. Exporting sugar at subsidised prices prevents other countries from exporting to those same markets: for instance, in 2001 Europe exported 782,356 tons of white sugar to Algeria and 152,000 tons to Nigeria, sugar Algeria and Nigeria would otherwise have bought from developing countries.

One country in particular that has lost out to Europe's sugar policy is Mozambique. It's one of the poorest countries in the world, decimated by sixteen years of civil war, which destroyed much of the country's agriculture and infrastructure. According to a report published by the Instituto Nacional Do Açúcar, 80 per cent of its people live in rural areas, and 70 per cent are below the poverty line. Mozambique has been attempting to rehabilitate its sugar industry since the war ended in 1992. Currently production costs are under £154 per ton, less than half of EU costs of £364 per ton. The industry employs 23,000 workers; if sugar mills were successfully restored the total number of jobs would rise to 40,000, and a further 10,000 jobs could be created in sugar-related sectors such as transport, packing, alcohol, paper and cattle-feed production.

Maria Gulela (not her real name), a widow with six children and two nieces (her brother's children who died of AIDs), is a typical sugar worker in Mozambique. For six months of the year she labours on the sugar plantation harvesting cane and earns £15 a month. Gulela says, 'With a fixed wage I manage to organise our day-to-day lives better and we have a guarantee that the children will go to school, that on a certain day of the month we will have sugar and soup at home.' She adds, 'The money, it is not enough.' The work is hard and conditions are poor: cane cutters labour for up to twelve hours a day cutting cane by hand. 'We work under the sun and rain without stopping. We have malaria problems because of mosquitoes . . . We asked the company to provide us with repellents but they said it would be too expensive and we should buy them with our own money. So we have to choose between avoiding mosquitoes and having enough clothes at home.' Sugar-cane workers like Gulela need better conditions, but she says, 'This job means that I am a human being because it brings me some hope in life.'

However, until last year the EU prevented Mozambique from exporting sugar to Europe, and, because of EU sugar dumping, its prospects have been curtailed. One of Gulela's fellow workers says, 'These policies are putting us at a disadvantage. They are rich and could give us a chance to live.' Another adds, 'Tell those people that our children are dying of hunger and disease. If our salaries could increase, life would be different.'

In 2001 the EU launched the 'Everything But Arms' (EBA) initiative, which would enable all products except weapons from the 49 least-developed countries to be sold to the EU without imposing any restrictions. Unfortunately, as far as sugar is concerned, implementation has been delayed due to heavy lobbying by the sugar industry. The quotas are still restricted and will only increase incrementally. In 2002 the total quota for EBA sugar was a mere 75,000 tons. Ethiopia and Mozambique are entitled to supply Europe with approximately 25,000 tons between them. This represents the equivalent of only a few hours' worth of European consumption – in other words, two of Africa's poorest countries have the right between them to supply Europe with sugar for just one day. All 49 countries in the initiative are – together – allowed to supply Europe with a total of three days' supply of sugar. This is the same amount produced by just fifteen sugar-beet farms in Norfolk. As a result, Oxfam estimates that

Mozambique, for example, lost the chance to earn an estimated £58 million between 2001 and 2004, almost three-quarters of the EU's annual development aid to the country.

In addition, the EU, as with ACP countries, is only allowing raw sugar to be imported into the country and is limiting refined sugar. Mozambique has sugar refineries and could make higher profits by refining raw sugar. Unrestricted access to the EU should take place by 2009, but this delay indicates that the EU is more concerned with shoring up its own agricultural output. Furthermore, Kevin Watkins, of Oxfam International, says, 'The EU chose to accommodate this increase in imports from least developed countries, not by cutting back on domestic quotas, but by transferring quotas from the ACP. This was a case of robbing the poor to give to the very poor.'

America is just as much to blame as Europe. It has a similar protectionist regime: high internal prices, a tariff rate of nearly 150 per cent and soft loans propping up domestic production. According to the Cato Institute's trade briefing paper on sugar, the US Government spends close to $1.68 billion buying and storing excess sugar to maintain artificially high domestic prices. America imports only 12 per cent of its sugar, leading the Cato Institute to estimate that developing countries lose out because they do not have access to the US market and the world price is thus lowered, costing the developing countries $1.5 billion a year. In the meantime, the 2002 US Farm Bill increased subsidies to American sugar farmers: they now receive $1.1 billion annually.

Poorer countries are affected in almost the same way as they are by Europe's sugar regime. For example, the Philippines is the twelfth-largest cane producer in the world. The country used to have a preferential trade agreement with the US but, when the US replaced much of the sugar in soft drinks with HFCS, the demand for sugar plummeted and imported sugar was cut by 70 per cent between 1982 and 1987. A quarter of a million people in the sugar industry in the Philippines lost their jobs. Many areas that are solely dependent on sugar-cane production, such as Negros Occidental, suffered from hunger and malnutrition. Farmers tried to grow subsistence food on former sugar estates that were no longer being cultivated, but were prevented by the sugar estates' private armies. In desperation many cleared rainforest but the land became unfertile in two to three years, forcing them to fell more trees, creating a series of environmental disasters.

Continuing low world-sugar prices still cause problems because it is more expensive to grow cane in the Philippines than it is to import cheap, subsidised American imports. According to official government figures, 41 per cent of people in the Philippines live below the poverty line. Permanent workers receive less than the official minimum wage and seasonal workers earn a third of the permanent workers' salary, as they are frequently paid by the piece.

The impact in the Dominican Republic was similar. A skilled cane cutter earns less than £38 a month, yet it costs almost £300 a month to feed a family of four. As the price of sugar has been forced down, the country has imported even cheaper labour from Haiti. The cane cutters live in compounds called *batays* next to the sugar-cane fields. Up to seven people live in a single room without water, sanitation or electricity. There is no organised union help and no healthcare.

Haiti was the first island in the Caribbean where people were forced into slavery; today Mark Ritchie, the director of the sustainable farming non-governmental organisation the Institute for Agriculture and Trade Policy says, 'The near-slavery conditions on plantations from Haiti to Negros in the Philippines are the wellsprings of exploitation and revolution and have been for hundreds of years. The social conditions for farmers and for the workers on these sugar-export factory farms are, literally, among the worse conditions found on this planet.'

What is absolutely insane about the whole sugar industry is that, even though we are eating enormous quantities of sugar, we are still producing more than we can consume and definitely more than we need. In 2000 world production was 136.8 million tons, an increase of more than 3 per cent on the previous year and a rise of 22 per cent over the previous 6 years. And yet world consumption was only 128.7 million tons. The EU alone produces more than five million tons of *surplus* sugar.

The EU's sugar policy will not be properly debated until 2006. Reform, including the full implementation of the EBA agreement, will help developing countries and could stop the ascendancy of the sugar beet. Michael Attfield, ex-Tate & Lyle director, says, 'I think the future is looking brighter. Life is going to become tougher for the beet producers.' The future may be bright for Tate & Lyle, as Attfield says, but Europe's former colonies will be hurt by any change to the artificially maintained ACP agreement.

For example, Fiji, like other ACP countries, receives a price for sugar from the EU that is double the world price and amounts to 84 per cent of its total export earnings. Fiji also has preferential access to the US market, where it sells 10 per cent of its sugar. A study published in the *Journal of South Pacific Agriculture* examined the implications for Fiji if European policy changes. The researchers concluded that, without any assistance from the EU, small farmers will suffer enormous losses. About a fifth of all households are already below the poverty line; this would rise to 80 per cent.

A coalition of aid organisations – Oxfam, Actionaid and CAFOD – have made suggestions for reforming the trade in sugar. They believe that Europe should eliminate export subsidies and dumping; reduce sugar production to below levels of European consumption; help ACP countries during the transitional period by raising aid money for this purpose through taxing sugar and the sugar industry; allow market access and finance to the least-developed countries; and, finally, support rural development and the environment.

Their guiding principle is that the burden of change should fall on those countries that are most able to deal with it. Overall they conclude that the emphasis for change should be in Europe, moving production in Europe away from beet.

Some countries have already taken the matter into their own hands. In parts of Kenya farmers are abandoning cane production and are growing food, supported by the non-governmental organisation Community Rehabilitation and Environmental Programme. Programme officer Thomas Barasa says, 'Farmers are better off producing food than growing a crop which earns them next to nothing.'

In the meantime, according to the *Guardian*, British Sugar and Tate & Lyle have argued that opening up the market would not help poor countries (only four of the ACP countries that do benefit are categorised by the World Bank as poor) but would allow in big businesses in Brazil and Australia.

In contrast, research commissioned by the EC shows that removing trade barriers to sugar would cut European production from just over 20 million to 6 million tons a year. Imports from the developing world are likely to rise to 10 million tons while European exports may be reduced to zero. Global sugar prices could rise by a third, helping the lives of millions of people around the world. Consumers would not lose out, however, because European prices are currently three times higher than the world average so prices of sugar on the shelf would presumably drop, and we would not be paying millions in our taxes to subsidise beet sugar.

Another option is that of fairly traded sugar. To comply with Fairtrade Foundation regulations, employers must pay decent wages, allow employees to join trade unions, provide housing where relevant and adhere to minimum health and safety as well as environmental standards. No forced or child labour is allowed. The Fairtrade organisation, for its part, guarantees a fair price for the product, but expects part of the profits to be invested in economic, environmental and social development.

The problem with Fairtrade at the moment is that, since the EU produces its own sugar, Fairtrade sugar is very expensive because it is competing against European sugar and has a large tariff slapped on it. Also, raw sugar needs to be processed immediately after harvesting and both mills and refineries are expensive. As a result, very few small producer co-operatives, such as those that exist for Fairtrade coffee or chocolate, can produce sugar of a high enough quality for export. In spite of this, fairly traded sugar began to be imported into Europe in the mid-1980s. The Fairtrade Labelling Organisation International started a certification scheme, and has thirteen sugar producers on its books, importing just over 650 tons of sugar a year. For example, Costa Rica has established a co-operative called Coopeagri, which has 5,000 members, 670 of whom cultivate sugar cane. It has its own refinery so can also produce white sugar and employs thirty permanent and ninety temporary members. Health insurance and social security is available to all employees. The amount exported on Fairtrade terms is small – about 160 tons – but the co-operative's total capacity at the time of writing is 5,600 tons of white sugar.

Barbados, once synonymous with sugar, is finding a third way. When Barbados was first colonised, the entire forest was felled to grow sugar cane. Now, ironically, scientists think that sugar cane is the only way to protect the island. Barbados is essentially a coral atoll; in some places the skin of soil is so thin, barely five centimetres, that it hardly covers the coral. But, when the cane is harvested, the leaves are left in the fields to act as a mulch that protects the soil from erosion. The other problem Barbados faces is lack of water – there are only two small rivers. Instead, rainwater percolates through the coral and collects beneath the atoll. It is then pumped back up for use as drinking water. As a result, the industry has to be careful of the amount of fertiliser it uses in case it ends up in the drinking water. As most tropical vegetables are given high doses of pesticides and herbicides, planting anything other than sugar cane is likely to increase both erosion and water pollution.

Since Barbados is one of the ACP countries, it stands to lose its quota in 2006 and it simply will not be able to compete. Already the sugar industry is running at a loss due to the country's high standard of living and relatively high wages (about $50 Barbadian – £15 a day for a cane cutter). Yet, to safeguard jobs and the island's environment, many feel they need to continue to grow sugar cane. So far the industry has come up with two novel ideas. The first is to breed wild cane that's known for its prodigious growth and use this cane, not for its sugar, but as fuel. A 24-megawatt electricity power station could be run most days of the year by burning 4,249.3 hectares of cane.

The second idea is to look at products within the cane itself. Using a new technique that doesn't crush the whole cane but removes the inner core and separates its components, other products could be harvested. Apart from sugar, sugar cane contains a remarkable number of chemicals, such as aconitic acid, a flavouring; succinic acid, which is added to clothes, ink, paints and could replace citric acid in drinks like Pepsi; and oxalic acid, used in textiles, detergents and pharmaceuticals. The pith, currently given to chicken farmers for litter, could be converted to wood or paper substitutes.

Environmental damage was first seen in Barbados as soon as the forests were cleared in the early seventeenth century. The rest of Europe's colonies in the Caribbean followed suit and, as the land was denuded for sugar cane, the soil quickly eroded and became relatively infertile. But the type of damage that is taking place now is of a different order of magnitude altogether.

Australia has been growing sugar cane since the nineteenth century, but it wasn't until the 1950s that scientists became aware of just how devastating sugar-cane production could be. The greatest damage has been to the Great Barrier Reef in Queensland. The reef is one of the richest natural systems on earth: it stretches for 2,000 kilometres along the east coast of Australia and includes almost 3,000 individual reefs, rainforests, mangrove forests and seagrass beds as well as 900 mini-islands and coral cays. Coral itself is complex; the reef is composed of more than 350 species, out of a total of 450 species worldwide. Each one has a symbiotic relationship with zooxanthellae, types of algae that live protected lives inside the coral skeleton, but make food for their host through photosynthesis. Humpback whales, dugongs and turtles breed in this region. Scientists are only just beginning to realise how complicated coral reef systems are.

Over the last 200 years, 80 per cent of the native vegetation alongside the reef has been removed for agriculture: in total, more than 50 per cent of Queensland's 117 million hectares of woody vegetation has been desecrated. Sugar-cane cultivation expanded rapidly: in 2004 400,000 hectares are in cultivation. The result is that soil is quickly eroded and runs into the wetlands, rivers, streams and ultimately the sea. In sugar-cane regions losses of 380 tons of soil per hectare have been recorded, compared to only four in natural rainforests. Since the 1950s fertiliser has been used in ever increasing amounts to keep sugar-cane yields high, and this too washes into waterways.

In 1985 there was an audit of Australian cane-growing practices commissioned by the association Canegrowers, which later in 1994 approved guidelines for sustainable cane growing. The sugar industry, initially reluctantly, moved to increase 'green cane' harvesting and 'trash blanketing'. Traditionally cane is burned before it is harvested to kill pests and make the fields more manageable for tilling after the harvest. 'Green cane', harvested without burning, is actually fresher and contains at least 6 per cent more sucrose. The 'trash', such as the leaves, is left on the soil, which helps improve soil structure and fertility and keeps the weeds down. At least 65 per cent of the crop is now harvested in this way. Over a five-year cycle of cane growth, fields that adopt these more environmentally friendly policies now only lose ten tons of soil per hectare on average per year. Nutrient loss, including that of fertiliser, is also reduced. Nevertheless, more than 20,000 tons of nitrates, a quarter of which comes from sugar-cane agriculture, still pour on to the Barrier Reef.

The impact is immense: in 2003 a report by an independent panel of experts commissioned by the Queensland Government showed that runoff from sugar-cane plantations was the main cause of the decline of up to 60 per cent of coral species in the inner section of the Great Barrier Reef. The high levels of nutrients from the fertiliser that wash into the sea promote the growth of plankton, which supports larger numbers of filter feeders, such as tubeworms and sponges, and these animals compete with coral for space. Algae also blooms rapidly in this newly nutrient-rich water, growing over coral and outcompeting fledgling coral colonies. In addition, high levels of nutrients have a two-pronged effect directly on coral – as well as killing established coral, they inhibit fertilisation rates and embryo formation so new corals do not seed. Phosphorus, also found in fertiliser, weakens coral skeletons, making them more susceptible to storm damage. When two areas in the Great Barrier Reef were compared – one that was undisturbed, and one that suffered from runoff – the latter had lower levels of coral biodiversity, smaller numbers of corals establishing themselves on the reef and a worse proportion of coral species.

The soil itself smothers and buries coral, reducing light availability, which prevents coral from photosynthesising properly. This inhibits growth and reproduction. In the Great Barrier Reef there are 5,000 square kilometres of seagrass habitat. Often referred to as the 'nursery of the sea', it provides shelter for juvenile crabs, prawns and fish: baby brown tiger prawns can live nowhere else. Dugong and turtles graze on this underwater meadow; seagrass beds also stabilise sediment with their root system. But excess sediment reduces the seagrass's ability to photosynthesise and the increased algae and plankton growth further reduces the amount of light getting to the plants.

Research by the James Cook University shows that dugong numbers in the Barrier Reef have plummeted by 50 per cent; other species, such as the Indo-Pacific humpback dolphin and the Irrawaddy River dolphin, may not survive into the next century. Mangrove forests are also being cleared in Australia – these forests are areas of high biodiversity, but 70 per cent of all inshore fish species are also dependent on the mangroves at some stage in their life cycle. Jeremy Smith, deputy editor of the Ecologist, says, 'It is a perfect, yet unbearably sad image of how connected the world has become. The production of a tiny, white grain that dissolves to nothing in a cup of hot tea is destroying the largest living organism on the planet.'

Runoff of sediment and fertiliser is not the only problem that sugarcane cultivation has wrought on the environment. Earlier in the twentieth century, Australian cane growers made a misjudgement. At the time, it must have seemed like a good idea, but it still has devastating effects today. Sugar cane in Australia was being ravaged by the larvae of two beetles, French's cane beetle and the greyback cane beetle. It was alleged that they had a natural predator, the cane toad,

Bufo marinus, which is found in South America and southern North America. So, in June 1935, the Australian Bureau of Sugar Experimental Stations imported a hundred toads from Hawaii to the Meringa Experimental Station near Cairns. The toads bred rapidly and, a month later, 3,000 were released. Not everyone believed this was such a good idea: former New South Wales government entomologist W.W. Froggatt and an Australian museum curator, Roy Kinghorn, protested. Releases were halted temporarily but commenced again in 1936. Within six months, 60,000 young toads had been let loose on Queensland.

Cane toads are the bullfighters of the amphibian world, heavy-set creatures with dry, warty, olive-brown skin. They typically grow to 15cm long, but the largest female seen was 24cm and weighed 1.3 kilos. Males are smaller and when they are ready to breed they develop 'nuptial pads', fleshy excrescences on their first two fingers that help them cling to the female, and entice females with a long, loud purring trill. Once they've mated, females can lay up to 35,000 eggs twice a year. Though captive toads can live for fifteen years, most reach five years. Virtually indestructible, they can survive losing half of their bodily fluids and can live in temperatures ranging from 5 to 40°C. They will eat practically anything – the tadpoles have five rows of teeth and feed on algae and other aquatic plants. The adults eat beetles, honey bees, termites, crickets and other insects, as well as other toads, frogs, snakes, small mammals, carrion, household scraps and pet food.

Even more alarming is the fact that, at every stage of their life cycle, cane toads are deadly poisonous. Although there have been no mortalities in Australia, they have caused temporary blindness, vomiting, breathing difficulties and inflammation of the nose and mouth; people who've eaten the toads or made soup from their eggs in other countries have died. The venom is produced by special glands and either oozes out, or can be squirted in a fine spray at the victim, causing a heart attack within fifteen minutes. In Hawaii, fifty dogs die every year from eating cane toads and in Australia the native wildlife, from crocodiles and dingos to tiger snakes, have been poisoned.

Not only do the toads breed rapidly, they also spread quickly, moving at a rate of five kilometres a year, and have almost no natural predators. They eat an inordinate number of honey bees, which is a problem for bee keepers and farmers who need their crops pollinated by the bees, and they either eat native fauna or compete with them for the same food, as well as transmitting diseases to native frogs and fishes.

As for the beetle grubs, the toads did eat them when they found them, but had no impact on the beetle population at all. Within five years of the toads' release, the cane growers discovered an insecticidal spray that worked on the beetle population and lost interest in the toads. So far, Australia has not found a way to get rid of them.

Shortly after Australia's toad debacle, America embarked on a course that has also had serious environmental repercussions and could be comparable to the devastation of the Great Barrier Reef. The Florida Everglades was a vast swampland radiating from Lake Okeechobee, one of America's largest freshwater lakes. It supplies Florida with fresh water, and the Everglades themselves were home to numerous species, from West Indian manatees to wood storks, Cape Sable seaside sparrows to snail kites. In some places the water is only fifteen centimetres deep, a glassy sheet of grass and water that has led to its description as a 'River of Grass'. At the turn of the twentieth century, it was seen as a wilderness to be tamed and drainage began.

The turning point came in 1947 when Florida was hit by two hurricanes and flood controls were insufficient to stem the deluge. Florida was devastated: the damage was estimated at £32.5 million. Between 1948 and 1971 the government embarked on a massive programme of water control; the Army Corp of Engineers fitted a vast system of pumps, dykes and 2,253 kilometres of straight-sided canals that animals could not climb out of and wading birds could not feed in.

The Everglades Agricultural Area (EAA) was created out of this reclaimed land, 80 per cent of which is used to cultivate sugar cane in the rich, moist soil. At 283,000 hectares it is the largest area set aside for farming in the world. Subsidies for sugar-cane production and tempting loan packages have encouraged US farmers to convert to sugar-cane cultivation. Now more than half of America's cane is grown in the Everglades. In the meantime, environmentalists, led by landscape architect Ernest Coe, campaigned to leave a small area untouched.

It took more than a decade for the first core area to be designated a national park. Now the Everglades National Park is the largest designated wilderness in America: 610,478 hectares containing more than 93,100 hectares of mangrove forest, and the most extensive mangrove ecosystem in the western hemisphere. It is recognised as being the most significant breeding ground for tropical wading birds in North America. It has recently been designated a World Heritage Site, an International Biosphere Reserve and a Wetland of International Importance.

Unfortunately, it is sited just below the EAA.

Although old-timers complained that bird populations had declined, it wasn't until 1981 that serious questions started being asked. Burkett Neely is the manager of Loxohatchee National Wildlife Refuge. He was concerned that the park was full of cattail. Initially this doesn't look like a problem: cattail, after all, is a native grasslike species that grows in the swampy water of the Everglades. But what Neely was seeing was 323,752 hectares of solid cattail growth in his reserve. Such heavy cattail growth prevents other species from surviving and stops wading birds from using the waterways. Neely started collecting data and discovered that vast amounts of phosphorus were present in the Refuge. As in the Great Barrier Reef, runoff from the plantations, laden with nitrates and phosphorus, was pouring into both the Loxohatchee reserve and the Everglades National Park and literally choking the wildlife. When flood warnings are issued, there is no longer enough land in the Everglades to absorb the impact. So, unlike Australian sugar-cane cultivation, the situation has been exacerbated by the proliferation of the canals and pumping systems. Huge water stations pump phosphorus-rich water into the canals and national parks. One of the biggest, S5a, has six pumps and can pump 1,659,322.85 litres a minute – enough to fill a swimming pool in one second.

In less than fifty years, more than half of the Florida Everglades has been drained. There are now fewer than thirty Florida panthers – a reddish-brown cougar endemic to southwest Florida – left; in fact, there are more statues of the big cat than there are real ones. Fifty-six species are either on the endangered or threatened list, the highest number for any state in the country. Since 1930 there has been a 90 per cent decrease in the number of wading birds, such as white ibis and roseate spoonbills. During the past thirty years, all animal species have declined by between 75 and 90 per cent.

Biologist Dr John Ogden, from South Florida Water Management District, has noticed a fundamental change at the very heart of this complex ecosystem. One of the major components of the Everglades is a slimy substance coating the plants in the marshes. It is made up of a number of different types of freshwater algae and is food for many tiny freshwater creatures and fish that other larger animals feed upon: in other words, it is the basis of the whole food chain in the Everglades. But recently, the change in the water chemistry has altered the composition of these algal colonies and now the creatures that once fed upon it are no longer able to do so.

Neely, in collaboration with the US Fish and Wildlife Services, took a brave and unprecedented step: they sued Florida for ignoring their own pollution laws. Vietnam veteran, Dexter Lehtinen, was their lawyer. He says, 'Lawsuits for the Everglades have never been based on any of the new clean water act principles or anything else. They're based on common law principles of anti-pollution – the up-stream guy cannot dump on the down-stream guy.' The state of Florida went on the defensive, hiring a team of lawyers from Washington and spending $6 million to prove that not only was it not the culprit, but that there was no pollution.

In 1993, however, the governor of Florida surrendered to the federal government and said that Florida would now enforce its pollution laws against the sugar growers. And that is when Big Sugar stepped in, spending $11 million, to defend its interests. Chief executive of US Sugar Robert Buker says, 'We wanted to merely come in and defend ourselves because it was a lawsuit that was blaming us and in which we were not allowed initially, so we had no choice but to file whatever litigation we could to try and get the facts heard.' The biggest of the Big Sugar families are the Fanjuls, Alfie and Pepe, originally refugees from Cuba and now the directors of Florida Crystals Corp. Their personal fortune is conservatively estimated at $500 million. Lehtinen says, 'In many ways I've always had a certain grudging respect for the Fanjuls, who are able to successfully use a system in which weaker policy makers and weaker individuals cave in to this more powerful, strong-willed family. You have to respect that, but you sure don't respect the policy makers who willy-nilly go along with this kind of thing.'

According to Edwin Moure, Everglades Campaigner with the National Audobon Society, 'In 1993, the Fanjul sugar family began direct negotiations with the Secretary of the Interior. It's important to note that an awful lot of the people who care about the Everglades were not party to these negotiations.' As a result of these talks, the 1994 Everglades Forever Act was formulated. At first glance, it looks positive: sugar farmers are required to put money directly into cleaning up the Everglades and restoring it – some $230 million. However, the total bill will be $8 billion. That's fair, says Robert Buker, 'The water that goes into the Everglades doesn't just come off the sugar farms. Some is from Lake Okeechobee; some is from West Palm beach. We are paying the amount needed to clean our own water and the rest of the cost is being carried by general taxes because the water comes from someone else.'

Others disagree. Lehtinen says that runoff does indeed filter into the Everglades from urban areas and highways, but he adds, 'Look at all these areas and the conclusion you'll come to is that sugar is the dominant, overwhelming problem.' In addition, the act stipulated that existing levels of phosphorus could continue to flow into the Everglades until 2003; thus reduction of the level of this chemical has only just started to take place, leading some campaigners to dub it the 'Everglades Whenever Act'.

Environmentalists claim that the only way to restore the Everglades is to cut the subsidies to the sugar-cane growers. Research by the government organisation the General Accounting Office estimates that, in the absence of any subsidies, 70 per cent of the industry would still survive. Moure suggests that, if a fifth of the land planted with sugar cane were replaced with a 'flow wave' – a wide path of water sweeping from Lake Okeechobee to the coast, the Everglades could be restored. Naturally, the sugar industry does not agree. Comments Lehtinen, 'The sugar companies are yelling, "The sky is falling, the sky is falling," to preserve their incredible profit levels.' Halfway through 2003, the Governor of Florida, Jeb Bush, passed a law that relaxed requirements to clean up the Everglades. Essentially the bill allows water-quality standards to be reduced, which means levels of phosphorus do not have to be dramatically lowered. The deadline for reduction of phosphorus – to the level that was established as a standard back in the 1990s – is now 2016.

Sugar cane is a problem. The selfish and absurd policies generated by developed countries are causing misery and poverty for millions. Reliance on sugar cane has led to widespread environmental destruction, since it is grown in vast monocultures with little species diversity. Land that was once forest or swampland is cleared to make way for its expansion, resulting in soil erosion, sediment and fertiliser runoff, which pollutes reefs and waterways. Herbicides and pesticides leave residues of heavy metal in the soil as well as poisoning wildlife and adversely affecting sugar-cane workers' health.

Sugar from sugar beet is little better. Like sugar cane it is grown in a monoculture creating a sterile environment for wildlife. Beet is also fed heavily with fertiliser and pesticides. Beet farmers use an average of 10.5 active herbicide ingredients per year, more than twice as much as is used on any other crop. According to the Soil Association, these practices particularly affect birds: between 1987 and 1998 the numbers of breeding lapwings declined by 50 per cent in the UK. Water requirements are three times higher for beet than for cane and the crop is one of the major causes of soil erosion in the UK – more than 350,000 tons are lost every time beet is harvested.

The obvious solution would be for beet to be farmed organically, but the sugar industry is concerned that this would cause a loss of profits. Instead they are turning to science.

12. INTO PRIMEVAL PAPUA BY SEAPLANE

Wouldn't primitive man be astonished if he could see us now, with our radio, our high-power rifles, and flying machines! For the bald truth is that neolithic man has not vanished entirely. In certain nooks and crannies of the world he lives now just as primitively as he did uncounted thousands of years ago. I found him just so when I alighted from a plane before a cannibal camp in the remote jungles of New Guinea and was mistaken for a god!

E.W. Brandes, *National Geographic Magazine*, 1929

Dr E.W. Brandes, an American plant pathologist, and Jacob Jeswiet, an expert on sugar-cane cultivation, visited Papua New Guinea in 1928. They knew of the wild type of cane, *Saccharum spontaneum*, but in fact, there was another main type, which Jeswiet discovered. He found *Saccharum robustum*, a vigorous, grassy cane used for thatching houses and fencing, growing along the shores of the Sepik and the Laloki Rivers. At the time, Brandes remarked firmly:

The idea of independent origin of the sugar cane in places other than the natural habitat of *S. spontaneum* and *S. robustum* is corroborated neither by present botanical evidence, nor by reports and writings that will stand scrutiny, and it is safe to assume that the cradle of the sugar cane is the region where the two wild species are found.

This discovery of a second type of wild cane in New Guinea could help to explain the mysterious origin of the Creole cane. One theory is that some *robustum* plants were slightly sweeter than others and the Polynesians selected these ones to grow in their gardens as a cane to suck upon. Over time, therefore, *robustum* evolved to become this new variety, the Creole cane.

Back in 1888, John Redman Bovell had revolutionised sugar-cane production with his discovery that the plant could flower; he saved the industry through his arduous and exhaustive breeding of the old cane species. Ultimately, though, breeding within one variety meant that sooner or later Bovell would run out of new genes and novel genetic combinations. Fortunately, a fresh hybrid was created just in time. At the Dutch cane-breeding station in Java, Proefstation Oost Java, some wild cane growing alongside the institution bred with one of their Creole canes. Because wild cane is so vigorous, this accidental off-spring was better at ratooning (regrowing after being cut back). Creole canes do not ratoon well – they can only be cut down two to three times before their yield declines dramatically, but these hybrids could keep going for three or four crops. The research institute, realising what a find they had, embarked on a breeding programme: one of their most popular canes, named after the station (POJ 2878) was exported to all the cane-growing regions of the world.

The US Department of Agriculture, when they finally realised how limited the gene pool was, sent Brandes, Jeswiet and C.E. Pemberton, a 'seasoned explorer', in search of parasites of sugar-cane pests, to Papua New Guinea 'to seek a disease-resisting sugar cane to revive a sick industry in our own Southern States'. So, although Brandes and Jeswiet's expedition would eventually end up confirming the roots of sugar cane's history, it was designed rather as a desperate attempt to save an ailing sugar industry and was the start of what would become a world-famous scientific sugar-cane breeding project.

Brandes, Jeswiet and Pemberton were accompanied by Dick Peck, a 32-year-old pilot who had that rugged, all-American glow about him. Brandes credited much of the trip's success to Peck's skill: during the expedition, which lasted 200 days, he made 57 flights of 16,000km in a small seaplane. On one trip they headed inland from Port Moresby, but the fog was so thick they could not even see the plane's wings. At 4,267m it started to rain. Brandes said:

> I saw Peck look anxiously at the thermometer on a wing strut. It registered 33 degrees Fahrenheit [just above zero degrees Celsius – the freezing point of water]. A drop of one degree would mean taking a load of ice on the wings. We could not lose altitude without danger of crashing into the side of one of the numerous mist-shrouded peaks just below. Fortunately, after twenty minutes of suspense, we burst into bright sunshine.

Another time, the three landed in a river with a strong current and were swept downstream. Brandes managed to halt their progress with an anchor rope, but then they swung crazily from one bank of the river to the other, and each time the wings missed being crushed against the trees by a hair's breadth. The anchor rope started to sing under the strain and was on the point of snapping. Then they realised that they were being watched by one of the indigenous people, the only man they had encountered on their entire trip who had not run away from them at first sight. He saved them by seizing one of their ropes and securing it to a tree trunk. He went on to procure food and drink for them, Brandes's comment was, 'We decided that he was a mental case, but an extremely useful and accommodating one.'

A few years before their trip, Peck had built a house for his wife and family, but she died in childbirth and he never returned to their home. He used to say to his friends, 'It's an adventurous life and I suppose that some day they will pick me out of a crash.' In 1931 he was the co-pilot of a plane owned by the *Chicago Daily News* called the *Blue Streak*. He was training to break the record for carrying a 5,000 kilos load over a 2,000km course when one of the wings disintegrated. The plane smashed into an oat field; there were no survivors.

Papua New Guinea is startlingly beautiful, surrounded by coral reefs and palm-fringed beaches, mangroves and sago palm swamps, sheer-sided mountains with jungle clinging to their slopes, their tips wreathed in mist. Once in the 'dark, evil exterior', full of 'blood-thirsty tribes', Brandes and Jeswiet knew immediately they had come to the right place, for the natives made spears out of wild sugar cane. Brandes spent some time observing them and describing their habits: the first tribe he met was polygamous and he noted that they had a *ravi*, or men's clubhouse for bachelors, but, he added, 'The married men also repair to the *ravi* at intervals, for the sake of a little peace and quiet.'

The team then 'took off . . . in earnest to start our search for sugar cane in the untamed regions where it and almost everything else were reputed to be very wild'. Brandes says of the first people in the interior that they managed to frighten with their plane, 'It goes without saying that we were considered supernatural.' At the first village they stopped at Pemberton found a length of chewed cane on the beach and made signs that they wanted food. The 'savages' immediately stopped waving their spears at them and brought them armfuls of cane, which they traded with the scientists. Brandes was excited to note that there were six different varieties in the bundle they had purchased. The natives then tried to sell the explorers stuffed human heads in return for fishhooks, razor blades, cloth and cigarette tins.

Brandes preferred the pygmies who lived in tree houses fifteen metres above the ground. Their houses were surrounded by saplings cut so that they hovered fifty centimetres above the path and a metre apart. He said that the pygmies bounced along on these obstructions like squirrels or cats, but under Brandes's rather thickset frame, they snapped. He said, 'I could not conceive of a more effective way for the pygmies to confuse and confound their heavier pursuers until they were able to assemble for defence of their little fortresses.' The pygmies gave them sugar cane, but bartered ornaments and weapons in exchange for matches and beads. Brandes and his colleagues had begun the trip taking tea and biscuits in true white-man's fashion; by the end they were eating cassowary sausages, roast hornbill and crocodile eggs.

Further up the Sepik River the team discovered a cane that was not sweet but was extremely vigorous with deep, garnet-red flesh.

When they returned to their boat they found that their cane cuttings had already started to sprout and they were forced to create a garden on the deck. Altogether they managed to bring back 130 different varieties. Brandes said:

It is conceivable that some of these varieties, propagated on a commercial scale, will eventually reach proportions gigantic in comparison with the limited amounts found in their native habitat. Thus we see that races of plants, like races of people, may migrate from one far part of the world to another to multiply and replenish the earth.

In spite of Brandes and Jeswiet's heroic expedition, by the 1960s it became apparent that the new varieties of sugar cane were not much better than the old ones. The sugar-cane industry had well and truly run out of genes. In the late 1960s it was decided that a genetic base-broadening programme should be launched in Barbados. As Brandes and Jeswiet had done, scientists would return to Papua New Guinea, as well as travelling to India and Indonesia, in search of wild cane species, particularly *Saccharum spontaneum*. These new varieties would be systematically bred with each other and the resulting off-spring then crossed with existing types of cane in order to introduce disease resistance and vigour from the wild canes, but retain the commercial canes' prodigious sugar-producing ability.

Dr Seshagiri Rao, a cytogeneticist originally from southeast India, went to Barbados in 1968 to work on the programme and ended up as director of the West Indies Central Sugar Cane Breeding Station (WICSCBS) in 1988. Rao's office, overshadowed by a giant frangipani tree, is at the top of a hill; like a shaken-out cloth, the experimental sugar-cane fields fall away in gentle undulations; steam from the nearest sugar-cane factory spreads across the horizon, blurring into the blue of the ocean. When the sugar cane flowers, there is a sea of cream plumes as far as he can see, flowing in waves formed by the wind. As if to remind the research staff why they are there, outside Rao's office are three concrete containers, like giant window boxes, bursting with *S. spontaneum*, a truly overgrown grass several metres high, desperate to escape its confines and rampage invasively through the experimental plots.

'We were the forerunners,' says Rao. 'We have the biggest collection of hybrids from wild canes and commercial varieties in the world.' The cane breeding station has amassed more than 3,000 varieties to date, 70 of which are new *S. spontaneum* clones. However, it was an arduous undertaking; the first commercial varieties, B 79474 and B 80251, took twenty years to develop.

The person who does much of the hard work is Dr Sandra Bellamy of the next-door Agronomy Research and Variety Testing Unit. Her father was a sugar-cane grower and she says, 'It's in my blood.' She is a native Barbadian and her office is on the top floor of an old plantation house with windows on three sides looking on to the cane fields. Every year 200 sugar-cane crosses are made from 120 different varieties resulting in 20,000 seedlings. Bellamy is responsible for the care and propagation of these seedlings.

The seedlings, potted in soft-drink bottles, are 7.5cm high within three weeks; in twelve months they reach three metres. Over the years she inspects them visually for a number of different traits, such as disease resistance, vigour and how upright they are, before beginning more quantitative work – measuring their sugar, fibre content and weight. The experimental strips are separated with Ng 2813, a distinctive variety of cane that, with its scarlet stalks and purple leaves, acts as a biodegradable marker. Bellamy selects the best canes for continued cultivation until finally, twelve to thirteen years later, she may have between one and six new varieties. Initially she used to say to the farmers that the Testing Unit had developed some new canes, which they were welcome to take. There were never any takers. 'Now I say to the men, "I have three varieties, can you come and give me your opinion?" And that works.'

It was only in 2000 that Bellamy saw the first variety she had developed growing in a field. 'It's frustrating that it takes so long, but a tremendous feeling.' She adds that she still wants to select 'a super-duper' sugar cane.

In spite of all these efforts – and even though canes have improved – the sugar content has remained the same for the past forty years. Many thought that this was simply a physiological limitation of the plant. But one man questioned this assumption.

Dr Anthony Kennedy, originally from England, turned up at the WICSCBS after completing his PhD on the genetics of foxgloves. Several years later he returned. 'I had the same job as I'm doing now essentially,' he says. 'I'm a geneticist and a plant breeder.' Kennedy, a small, shy, somewhat diffident man, festooned with a huge beard, started to wonder why sugar content had not increased. Currently it takes about nine tons of sugar cane to produce one ton of sugar. Yield, the amount of sugar cane that is grown, is extremely environmentally dependent. For instance, if it doesn't rain, the cane will grow poorly. Sugar content itself, however, is genetically heritable and is not influenced by environmental factors to the same extent but has remained at 19 per cent or below. Why, Kennedy wondered, had sugar levels but not yield remained the same?

He didn't believe that cane was at its physiological limit, but suspected that breeders must be inadvertently selecting for other traits rather than concentrating on sugar content, which, after all, is why sugar cane is grown in the first place. He says:

I made the argument that sucrose content is the most important thing because every ton of cane costs you money. You've got to cut it, harvest it, transport it to the factory; the factory has to grind all that mass of cane and squeeze the juice out, then it's got to use energy to boil the juice out of the sugar. If you put more and more sugar in every cane, even if the yield is not as good, economically it's much better because it doesn't cost you any more money to harvest but your returns are better.

In the 1980s, Kennedy embarked on a programme that some considered foolhardy.

Opposite the WICSCBS is an immensely tall corrugated-iron shed. Kennedy picked canes in bloom, like cut flowers, and put them in buckets of a special solution with food, and antifungal and antibacterial properties to keep them alive, and then placed a large paper lantern round each male and female pair. He made sure that everyone working at the station took it in turns each morning to shake the males so that the pollen would fall upon the females. He then bred the resulting seedlings and tested them for their sugar content. 'If it had a high sugar content,' he says, 'whether or not it looked funny, I said I wanted it to be another parent.' As each generation is grown for three years it's a long process – and after the second cross it became apparent that the sugar content had not shifted. A number of Kennedy's detractors suggested that he should stop, but he continued . . . and a strange thing happened. 'It was really quite spectacular,' says Kennedy.

Using a sharp metal tube, Kennedy punches into the cane stem and extracts some of the plant's juice. Then with a handheld refractometer he measures Brix, which is the total dissolved solids within the juice. Of course, apart from sucrose there will also be minerals, waxes, resins and so on, but, in general, Brix correlates well with total sucrose content. Kennedy had decided that he would tag all those canes that contained a certain level of sucrose, until he found that he was tagging almost all the canes in the field. He raised the level of sucrose twice more as he realised, with increasing confidence, that his experiment had finally worked.

Now on the fourth cycle of breeding, some of these high-sugar-content canes have been bred with commercial varieties so that they have all the other desirable traits growers need, such as disease resistance, as well as high sucrose levels. What Kennedy thinks happened is that, in the past, farmers threw away high-sucrose types because they were not paid for sugar, they were paid by the ton. 'There was no pressure, in other words, to select high sucrose content,' he says. Kennedy's research has increased sugar content to 23 per cent, which will have a considerable impact economically. 'That was a big part of my life,' he says, 'and I'm very excited because no one else in the world has seen sugar contents of the kind that we're getting. They're a world record.'

Interestingly, through this research, Kennedy has learned more about the way that sugar cane actually produces and stores sugar. 'It's not just a passive thing,' he says, 'there's an active process of storing sugar and keeping it in the cell. We've changed that process so that it will allow more and more sugar to be stored.' Kennedy doesn't believe he's reached the plant's limit – but there will be one. What will curtail sugar content is simply the fact that the cells will kill themselves. Sugar is a preservative, which is why it works well in jam-making; it forces out water and creates a medium that fungi and bacteria – and plant cells – cannot survive in.

However, not all scientists think that Kennedy's laborious, old-fashioned approach is the way forward. Researchers at the University of Stellenbosch in South Africa analyse the way in which sucrose accumulates in sugar cane and recommend specific enzymes that could be targeted for genetic manipulation. Cuba's decaying industry may be wiped out by the 2006 EU reform; they think their only hope of saving sugar cane is to genetically engineer the plant so that it produces fructose instead, which will earn them twice as much as sucrose does. But the only country that has already started growing outdoor trials of genetically engineered sugar cane is Australia.

Sugar is Australia's second-largest export after wheat. It generates almost two billion dollars for the economy and employs more than 17,000 people. The Australian sugar industry believes that genetic engineering is vital to maintain and enhance their competitiveness on the world market so scientists in Australia are currently attempting to increase sugar content. However, they are having more success in another field: making white sugar that doesn't have to be refined.

When sugar cane is crushed in the mill, an enzyme turns the sugar brown – it's the brown colour that refineries spend millions trying to remove. At the Commonwealth Scientific and Industrial Research Organisation (CSIRO) in Queensland, a team led by Dr Chris Gof have found the gene that codes for this enzyme. In order to trick the plant, they've taken out the gene and then replaced it, but back to front. Although this is, strictly speaking, genetic manipulation, the gene is the plant's own. Sugar cane with the backwards gene is being grown in giant glasshouses but has not yet been released for field trials or to commercial growers. But if this variety succeeds, it could spell an end to sugar refining.

What is more controversial is a genetically manipulated cane currently undergoing field trials that contains genes from two different bacteria as well as a jellyfish. The jellyfish, *Aequorea victoria*, contains a green fluorescent protein (GFP). The gene that codes for this protein has been inserted into sugar cane to act as a marker – to help the scientists tell whether other genes inserted into the plant are switched on and are working properly or if undesirable genes have been turned off. When researchers shine an ultraviolet light on the plant cells and look at them through a microscope, they can see the GFP glowing. 'It's a particularly good marker gene,' says Dr Adrian Elliott, who works with Gof, 'because it can be easily switched on – and you can see it working at once.' He adds, 'The GFP is strictly a scientific tool. You wouldn't include it in any crop for human consumption, simply because there is no need to. Even if it was, sugar is a pure, refined product and we would not eat the protein or gene, which is disposed of with the sugar cane during refining.'

Elliott has also attached the GFP to a gene from a microbe, agrobacterium, which encodes for antibiotic resistance. A third gene from another type of bacteria has been added but is not actually active in the sugar cane. This genetically manipulated cane is a test in every sense of the word – it has been created through tissue culture in the laboratory in order is to see whether GM cane per se (no matter where the added genes are from) grows differently from normal cane.

However, some scientists are concerned: Carlo Leifert, professor of Ecological Agriculture at Aberdeen University in Scotland, finds the work 'incredibly worrying'. Leifert says:

If you add an alien gene to a plant, how do you know what side effects you will get? It is another example of the completely unpredictable nature of genetic transformation. By putting a jellyfish gene into sugar cane you could change the spectrum in the plant – it could change the toxins the plant produces to protect itself from insects into toxins that could potentially harm humans.

In the short term it will prove difficult to genetically manipulate sugar content in sugar cane because the genetics of the plant are fiendishly complicated. Sugar from genetically modified sugar beet is likely to appear on our shelves far sooner.

With the changes to European subsidies and the reform of the Sugar Protocol in 2006 threatening profits and livelihoods, GM beet could be the answer – or so the beet industry thinks. Broom's Barn, the national centre for sugar-beet research in Suffolk, is conducting trials on beet that is herbicide resistant. It needs 80 per cent less weedkiller than other sugar beet, at a saving of £150 per hectare a year. This research could help reverse the trend of herbicide use by sugar-beet producers which is decimating wildlife. According to Broom's Barn's director, Dr John Pidgeon:

> Frequent spraying destroys the weeds on which the insects and birds feed, but our system means we can reduce the amount of spraying and allow weeds in between the rows to flourish in summer without affecting yield. We are very excited about our results because this is the first time research has shown that GM herbicide tolerant crops can be managed for environmental benefit.

He hopes that, if trials are approved, the GM sugar will be on sale by 2007.

What will help conventional breeding techniques as well as genetic engineering will be the unravelling of the sugar-cane genome. A number of other organisms have already had their genome mapped, including humans, yeast, the domestic silkworm and the Chinese cabbage. Sugar cane has one of the most complex set of genes found in a plant but, in spite of its importance, had not been tackled. A quarter of the world's sugar is produced by Brazil, so it is perhaps no surprise that they have co-ordinated the mapping of the sugar-cane genome.

In May 1999 the Sugar Cane EST Genome Project (Sucest) was launched by Professor Paulo Arruda of the Universidade Estadual de Campinas in São Paulo. Rather than logging DNA directly, teams of geneticists from around the world worked with Arruda to log ESTs – expressed sequence tags. 'If you did a complete sequence of DNA it would use lots of resources,' says Arruda, who explains that only 10 to 15 per cent of sugar cane's DNA is actually 'expressed' or used by the plant. In fact, because so much of its DNA is repeated and thought to be 'junk', many estimated that sugar cane would not have any more genes than a small and less complex plant, such as *Aribidopsis*, the botanists' version of the lab rat.

ESTs help researchers track only the DNA that is being used. What happens in the plant's cell is that functioning DNA is copied on to another molecule, mRNA. This molecule is, in effect, the mirror image of the original DNA and it is made and transported into the heart of the cell to be translated into a protein or an enzyme, or whatever the gene coded for. The researchers capture the mRNA and work backwards to recreate the original DNA sequence – then they make copies, or clones, of it. This, then, is an EST – the DNA that corresponds to a functioning mRNA molecule. Arruda aimed to find 300,000 ESTs that correlated with 50,000 genes by 2003. He and his international teams were successful.

But now the hard work will begin. No one knows what most of these genes do. The dream of tinkering with a few genes to boost sugar production is still a long way off: many genes are involved in increasing sugar. Scientists are also searching for genes that will help plants survive heat, cold, drought and pathogens.

So far the function of only one gene that Sucest identified has been found: this is a major gene in the sense that it controls a trait by itself without the input from other genes. It was discovered by Dr Angélique D'Hont, a geneticist based at the Centre de Coopération Internationale en Recherche Agronomique pour le Développement (CIRAD) in Montpellier, France, and her team. The gene is particularly exciting because it is responsible for resistance to what's been called 'rust'. Rust is a disease caused by a fungus, *Puccinia melanocephala*, which causes reddish-brown spots and lesions in the plant's leaves. It probably originated in India but is now capable of decimating sugar cane worldwide.

D'Hont is slim, blonde, boyish and slightly scruffy in her plant-breeder uniform of faded jeans and checked shirt. A keen horserider, rock climber and paraglider, she has just had a baby girl and has had to slow down a little physically if not mentally. A brilliant scientist, D'Hont has been responsible for many of the breakthroughs in sugar-cane research, including unravelling sugar cane's tangled history using some of the latest gene technology. Back in the 1990s she smoothed the path for Sucest by showing exactly how complex the sugar-cane genome really is.

D'Hont had gone to Australia to study a particular cross between two varieties of sugar cane, but, when she arrived, the cross had not worked. She asked her boss if she should come back to France, but she was recommended to stay on, learn English and find another research project. There was no team doing the same kind of work as D'Hont so she had to struggle by herself. It took her a year to learn the basic techniques she needed, and which she can now teach a student in a month.

Most organisms, such as humans, are diploid – we have two sets of chromosomes. The chromosomes contain the genes. During sexual reproduction, one set of twenty-one chromosomes from each parent join together to form a new, diploid person. Sugar cane is a polyploid – it has more than one set of chromosomes from each parent plant – which makes it far more of a headache for scientists.

What D'Hont did was work out exactly how many copies of each chromosome there are in sugar cane. She found that *Saccharum officinarum*, the Creole cane, has 10 chromosomes, but they are copied 8 times. *S. spontaneum* has 8 chromosomes copied in sets of 8, so its total number of chromosomes can range from 40 to 128 – in other words, it has between 8 and 16 copies of its chromosomes. *S. robustum* is large too – it has 40 to 80 chromosomes.

Polyploidy refers to the fact that the chromosomes are in multiples of a known number – eight in this case. But sugar cane, as D'Hont showed, can also be aneploid, with random numbers of copies – five copies of each chromosome, for instance. As Anthony Kennedy says, 'If Mendel [1822–84, sometimes referred to as the "father of genetics"] had used sugar cane instead of peas, he'd have gone mad. With eight copies of the chromosome the mixing up that can go on is phenomenal.'

It is now known why sugar cane is so tricky, but polyploidy does occur quite frequently in plants, even if not in quite such a devilishly complex manner. It could be sugar cane's way of hedging its bets. Sexual reproduction means each individual inherits a different set of chromosomes from its parents and some of these chromosomes may also mutate. As a result humans, for instance, are all unique and some of us will be better adapted to the world than others. Sugar cane usually produces vegetatively – its offspring are clones. By having lots of different copies of each gene within it – some of which could have mutated – a sugar cane often has several different versions of the same gene and, hopefully, at least one copy will help the cane adapt to its current environment. It is a little like a person having various different genes for eye colour. Suppose you are born into a family living near the Red Sea, you might switch off the gene for blue eyes and for green eyes, and switch on the one for brown eyes to help protect your retina from the sun's glare.

D'Hont published her work in 1998, a year before the sugar-cane genome project began. She then went on to show that sugar cane is genetically very similar to other important crops that are also grasses – maize, sorghum and rice. These species show synteny – the same genes from each species are on the same chromosomes – and coliniarity – their genes are arranged in the same order. Because rice and sorghum are economically important and relatively simple, their genomes have already been mapped. D'Hont argues that you could use the genetic map of sorghum, in particular, to help map sugar cane and figure out which genes do what by knowing what the analogous gene does in sorghum or rice. She's already found a number of genes that are involved in sugar production, but each one only controls 3–4 per cent of the total sucrose content, so research on manipulating yields this way still remains a long way off. D'Hont and her team at CIRAD are continuing to work on this map, which is an alternative and complementary approach to the Brazilian project. 'I think at the moment we have the best, the most saturated map, that exists on cultivated sugar cane,' she says.

D'Hont has also turned her attention to the past, to the origins of sugar cane. Many researchers believed that sugar cane did not evolve solely from the wild type, *S. robustum*, but had bred with other grasses, in particular a related, large, vigorous grass with tall, thick stalks called *Erianthus arundinaceus*. To try and exploit this vigour, breeders had crossed the two plants. But, when D'Hont examined the offspring, she found that a few of the chromosomes had been eliminated. This means the plant has fewer and fewer genetic possibilities as it evolves. Moreover, these hybrids were sterile. Therefore, *Erianthus* is highly unlikely to be related to the Creole cane. The main contributor was indeed *S. robustum*.

The next conundrum was the evolution of *S. barberi* and the Chinese cane *S. sinense*. Experts had a number of theories about their evolution, for instance, that they were a hybrid of *S. officinarum* and *Erianthus*. However, Brandes had predicted that *S. officinarum* migrated from New Guinea and bred with wild *S. spontaneum* canes in India to create *S. barberi*.

What D'Hont did was combine chromosomes from *S. barberi* with chromosomes from *S. officinarum*. Then she added dye that only stuck to one sort of chromosome: a brilliant green that attached itself to chromosomes from *S. officinarum*; and a livid red that dyed *S. spontaneum* genes. When she looked down the microscope, all the chromosomes were either red or green. In other words, no other varieties were involved in the formation of *S. barberi*. Brandes's prediction, made more than seventy years before, was correct: *S. barberi* is a hybrid of *S. spontaneum* and *S. officinarum* – India's wild cane and New Guinea's Creole cane.

D'Hont applied the same technique to dissect the components of a modern-day sugar cane. She found that there were red and green genes – 80 per cent from *S.officinarum* and 10 per cent from *S. spontaneum*. But 10 per cent were both red and green. Older varieties showed higher amounts of red-and-green genes. What this means is that those genes have recombined – one chromosome from *S. spontaneum* has bonded with one from *S. officinarum* to form a new version that cannot be split. It is an important finding, because, if there is a good trait on a recombined chromosome, say from *S. spontaneum*, it may be linked to a bad one, and it is impossible to have the good without the bad. It was the first time that plant breeders had realised that breeding *S. spontaneum* with *S. officinarum* would have this effect.

So, thanks to modern-day genetic techniques, we can finally trace the story of sugar cane. It was once a wild grass called *Saccharum robustum* that grew in New Guinea and was used by the local people as thatch for their houses and fencing for their gardens. Some *S. robustum*, through an accident of evolution, contained traces of sucrose. The local people selected these canes and grew them, chewing on the canes, as they still do today. Over time a sweeter version was created. Eight thousand years ago the New Guineans took this new cane – *Saccharum officinarum* – to India. In India it bred with their wild cane, *Saccharum spontaneum* to form *S. barberi*, a thinner cane with less sugar, but which was better adapted to adverse conditions. The Indians took *S. officinarum* to China, where it hybridised with their local *S. spontaneum* to form *S. sinense*, the Chinese cane. *S. officinarum* was then transported throughout the world and was called the Creole cane, one of a small number of varieties of *S. officinarum* referred to as the noble canes. The Creole cane then hybridised with *S. barberi* to produce the Bourbon cane that rapidly replaced it until it, too, was discovered to be susceptible to disease.

Other strains appeared – the Cheribon or Transparent series. John Redman Bovell created an extensive number of hybrids from within the noble-cane family. But the gene pool was practically stagnant. It was not until the 1920s that the modern version of sugar cane was created – a cross between *S. spontaneum* and *S. officinarum*, the off-spring of which was then bred back with *S. officinarum* to retain the wild cane's vigour and disease resistance, combined with the Creole's sugar-producing ability. This is what we essentially still grow today – but with the addition of new wild-cane varieties using more systematic and scientific breeding programmes. That is, until the genes no longer come from Indian or Indonesian cane but are derived instead from other animals and plants, whether they're jellyfish or bacteria, and are artificially inserted into a new, genetically enhanced version of sugar cane.

EPILOGUE

For me it was a special treat – a trip to London for the weekend. I got up early on Saturday morning and went to Borough Market where I was assailed by the heady, hazelnut aroma of roasting coffee and served a wake-up shot of single-estate Brazilian by a vendor whose hands shook. I bought blanched asparagus that was just coming into season, baby artichoke heads, their sepals flushed purple, and became stupidly excited at finding aubergines the colour and size of fat pearls.

And then I saw some sugar cane. It was peeled, cut into sections no bigger than wedges of baklava and blister-wrapped.

I have seen canes the colour of tourmaline tied in heavy bundles in terraces in Madeira; maple-leaf-red cane in an experimental plot in Barbados; stacks of yellow canes, brittle as bamboo, for sale in a market in Funchal; and pale-green segments, called chopstick cane, flying from the maw of a sugar-cane harvester. I have watched vats of sugar-cane juice stew in a dark, heat-soaked factory, pungent with its heady sap, and seen pure, white crystals spin through granulators as the hot, dry air absorbed any lingering moisture from within its molecular structure. It seems a peculiar irony to find that fresh, unadulterated cane had been packaged in plastic and flown thousands of miles to the very city that refines much of the world's sugar, to a market that once sold to the poor and now caters to the whims of the elite. As I looked at those segments of sugar cane, I found it hard to believe what it had been responsible for.

I have read accounts of African slavery that left me depressed and despondent at the true depths of human cruelty; I have struggled to understand the tortuous intricacies and Kafkaesque logic of European trade wars and their concomitant effect on the lives of hundreds of people in the developing world to whom even water and electricity is a luxury; I have failed to grasp why we need a chlorinated version of sugar at the expense of the lives of thousands of animals; and I have questioned the ethics of adding bits of bacteria and jellyfish to a plant. I have tried to cut down my own intake of sugar and failed to eliminate it from my diet: it is a strange and wonderful plant that in my case allows me to indulge in white wine, vegan chocolate cake and Booja

Booja midnight express truffles, which, I have discovered, I find it hard to completely do without. Those small slices of stem were from a strange and wonderful plant that in our hands has led to untold suffering and brought joy to the lives of millions.

BIBLIOGRAPHY

Anderson, E.N. and Anderson, M.L., 'Modern China: South' In K.C. Chang (Ed.), *Food in Chinese Culture: Anthropological and Historical Perspectives,* pp. 317–82, Yale University Press, New Haven and London, 1977.

Artschwager, E. and Brandes, E.W., *Sugar Cane: Origin, Classification, Characteristics, and Descriptions of Representative Clones,* US Dept of Agriculture Handbook No 122, Gov Printing Office, Washington DC, 1958.

Aykroyd, W.R., *Sweet Malefactor: Sugar, Slavery and Human Society,* Heinemann, London, 1967.

Ayrton, Elizabeth, *The Cookery of England,* Penguin, Harmondsworth, 1974.

Bacci, M., Miranda, V.F.O., Martins, V.G., Figueira, A.V.O., Lemos, M.V., Pereira, J.O., Marino, C.L., *A Search for the Markers of Sugarcane Evolution,* Genetics and Molecular Biology, 24(1–4): 169–74, 2001.

Barasi, Mary E., *Human Nutrition: A Health Perspective,* Second Edition, Arnold, London, 2003.

Barnes, A.C., *The Sugar Cane: Botany, Cultivation and Utilization,* Interscience Publishers, New York, 1964.

Barton, Laura, 'A Spoonful of Propaganda', *Guardian,* 12 April 2002.

Battersea and Wandsworth Trade Unions Council, *Sacked in Style by Tate & Lyle: The Story of Gartons, Battersea. The Death of a Factory,* Battersea and Wandsworth Trade Unions Council, London, 1983.

Bentham, J., *Leading Principles of a Constitutional Code,* London, 1823.

Billington, Roy Allen (Ed.), *A Free Negro in the Slave Era: The Journal of Charlotte L Fortin,* Collier Books, New York, 1961.

Blackburn, Frank, *Sugar-cane,* Longman, New York, 1984.

Blassingame, John W. (Ed.), Slave Testimony: Two Centuries of Letters, Speeches, Interviews and Autobiographies, Louisiana State University Press, Baton Rouge, 1977.

Blume, Helmut, *Geography of Sugar Cane: Environmental, Structural and Economical Aspects of Cane Sugar Production,* Verglag, Berlin, 1985.

Boseley, Sarah, 'Sugar Industry Threatens to Scupper WHO', *Guardian*, 21 April 2003.

Boseley, Sarah, 'US Accused of Sabotaging Obesity Strategy', *Guardian*, 16 January 2004.

Brandes, E.W., 'Into Primeval Papua by Seaplane', *National Geographic Magazine*, 56: 253–332, 1929.

Brand-Miller, Jennie, Foster-Powell, Kaye, Leeds, Anthony, Lintner, Lisa, *The Glucose Revolution: GI Plus*, Hodder and Stoughton, London, 2001 edition.

Brand-Miller, J.C., Holt, S.H.A., Pawlak, D.B., McMillan, J., 'Glycemic Index and Obesity', *American Journal of Clinical Nutrition*, 76: 281S–5S, 2002.

Briffa, John, 'Eating Lessons', *Observer Food Magazine*, January 2004.

Butterfield, M.K., D'Hont, A., Berding, N., 'The Sugarcane Genome: a Synthesis of Current Understanding, and Lessons for Breeding and Biotechnology', Proc Soc Afr Sugarcane Technol Ass 2001, 75: 1–5, 2001.

Catholic Institute for International Relations, *Sugar: Europe's Bittersweet Policies*, Catholic Institute for International Relations, London, 1994.

Chalmin, Philippe (trans. Erica E. Long-Michalke), *The Making of a Sugar Giant: Tate & Lyle 1859–1989*, Harwood Academic Publishers, Switzerland, 1990.

Child, Josiah, *New Discourse on Trade*, London, 1718.

Colagiuri, S., Brand-Miller, J., 'The "Carnivore Connection" – Evolutionary Aspects of Insulin Resistance', *European Journal of Clinical Nutrition*, 56(1): S30–S35, 2002.

Colantuoni, C., Rada, P., McCarthy, J., Patten, C., Avena, N.M., Chadeayne, A., Hoebel, B.G., 'Evidence that intermittent, excessive sugar intake causes endogenous opioid dependence', *Obes Res*, 10(6): 478–88, 2002.

Coupland, Reginald, *Wilberforce: A Narrative*, Collins, London, 1923.

Curtin, Philip D. (Ed.), *Africa Remembered: Narratives by West Africans from the Era of the Slave Trade*, University of Wisconsin Press, Madison, 1967.

Dalmey, Kath, 'Sugar and Spin', *Ecologist*, November 2003.

Darwin, Charles, *On the Origin of Species*, London, 1859.

Daugrois, J.H., Grivet, L., Roques, D., Hoarau, J.Y., Lombard, H., Glaszmann, J.C., D'Hont, A., 'A Putative Major Gene for Rust

Resistance Linked with an RFLP Marker in Sugarcane Cultivar R570', *Theor Appl Genet*, 92: 1059–64, 1996.

Deerr, Noël, *The History of Sugar*, Vol I, Chapman and Hall, London, 1949.

Deerr, Noël, *The History of Sugar*, Vol II, Chapman and Hall, London, 1950.

De Freitas Trean, María, *The Admiral and his Lady: Columbus and Filipa of Portugal*, Robert Speller and Sons, New York, 1989.

Denny, Charlotte, 'Sweet Smell of Cynicism', *Guardian*, 19 January 2004.

Dent, V.E., 'The Bacteriology of Dental Plaque from a Variety of Zoo-Maintained Mammalian Species', *Archs Oral Biol*, 24: 277–82, 1979.

Dent, V.E. and Marsh, P.D., 'Evidence for a Basic Plaque Microbial Community on the Tooth Surface in Animals', *Archs Oral Biol*, 26: 171–9, 1981.

Dewar, A., May, Mike J., Sands, Richard J., Qi, Aiming, Pidgeon, John D., 'A Novel Approach to the use of Genetically Modified Herbicide Tolerant Crops for Environmental Benefit', *Royal Soc Proc B*, 270(1513): 335–40, 2003.

D'Hont, A., Lu, Y.H., Feldmann, P., Glaszmann, J.C., 'Cytoplasmic Diversity in Sugar Cane Revealed by Heterologous Probes', *Sugar Cane*, 1: 12–15, 1993.

D'Hont, A., Lu, Y.H., Gonzàlez de Leòn, D., Grivet, L., Feldmann, P., Lanaud, C., Glaszmann, J.C., 'A Molecular Approach to Unraveling the Genetics of Sugarcane, a Complex Polyploid of the Andropogoneae Tribe', *Genome* 37: 222–30, 1994.

D'Hont, A., Rao, P., Feldmann, P., Grivet, L., Islam-Faridi, N., Taylor, P., Glaszmann J.C., 'Identification and Characterisation of Intergeneric Hybrids, *S. officinarum* X *Erianthus arundinaceus*, with Molecular Markers and in situ Hybridization', *Theor Appl Genet*, 91: 320–26, 1995.

D'Hont, A., Grivet, L., Feldmann, P., Rao, P., Berding, N., Glaszmann, J.C., 'Characterisation of the Double Genome Structure of Modern Sugarcane Cultivars (Saccharum spp.) by Molecular Cytogenetics', *Mol Gen Genet*, 250: 405–13, 1996.

D'Hont, A., Ison, D., Alix, K., Roux, C., Glaszmann, J.C., 'Determination of Basic Chromosome Numbers in the Genus Saccharum by Physical Mapping of Ribosomal RNA Genes', *Genome*, 41: 221–5, 1998.

D'Hont, A., Glaszmann, J.C., 'Sugarcane Genome Analysis with Molecular Markers, a First Decade of Research', *Proc Int Soc Sugarcane Technol*, 24: 556–9, 2001.

D'Hont, A., Paulet, F., Glaszmann, J.C., 'Oligoclonal Interspecific Origin of "North Indian" and "Chinese" Sugarcanes', *Chromosome Research*, 10: 253–62, 2002.

Donnelly, John, 'Bush Condemns Slavery as one of "Greatest Crimes": Speech at Source of African Trade gives no Apology', *The Globe*, 7 September 2003.

Drayton, Richard, *The Caribbean Mind Through the Prism of a Scientific Discovery: A History of the Emergence of Sugar Cane Breeding in Barbados 1859–1910*, Thesis, Harvard University, 1986.

Dresser, Madge and Giles, Sue, (Eds.), *Bristol and Transatlantic Slavery*, Bristol Museums and Art Gallery, Bristol, 2000.

Dresser, Madge, *Slavery Obscured: The Social History of the Slave Trade in an English Provincial Port*, Continuum, London, 2001.

Dufour, P., Deu, M., Grivet, L., D'Hont, A., Paulet, F., Bouet, A., Lanaud, C., Glaszmann, J.C., Hamon, P., 'Construction of a Composite Sorghum Genome Map and Comparison with Sugarcane, a Related Complex Polyploid', *Theor Appl Genet* 94: 409–18, 1997.

Dufty, William, *Sugar Blues*, Warner Books, USA, 1975.

Dunn, Richard S., *Sugar and Slaves: The Rise of the Planter Class in the English West Indies 1624–1713*, W.W. Norton and Company, New York, 1973.

Earle, F.S., *Sugar Cane and its Culture*, John Wiley and Sons, New York, 1928.

Eder, Steve, *Sugar Scandal*, Turner Original Prod Inc., Atlanta, 2002.

Edwards, Paul, (Ed.) *The Life of Olaudah Equiano*, Longman, Essex, 1989.

Elft, E.C., 'Adventures of Elgin's own daring aviator, Dick Peck', *Courier News*, 23 February, 2004.

Elliott, S.S., Keim, N.L., Stern, J.S., Teff, K., Havel, 'Fructose, Weight gain, and the Insulin Resistance Syndrome', *American Journal of Clinical Nutrition*, 76(5): 911–22, 2002.

Engel, Astrid, 'Sugar: Bitter rather than Sweet', *Fair Trade Yearbook*, London, 2001.

Engler, Mark, 'Cattail Country', *New Internationalist*, 363: 23–4, 2003.

Fa Hien, *A Record of Buddhist Kingdoms*, 337–422.

Fairrie, Geoffrey, *The Sugar Refining Families of Great Britain*, Tate & Lyle, London and Liverpool, 1951.

Fearnley-Whittingstall, Hugh, 'A Fat Lot of Good', *Observer Food Magazine*, January 2004.

Finn, J.P. and Lord, G.H., 'Neurotoxicity Studies on Sucralose and its Hydrolysis Products with Special Reference to Histopathologic and Ultrastructural Changes', *Food Chemical Toxicology*, 38 (Suppl 2): S7–S17, 2000.

Frances, Karen, Hird, Vicki, Lobstein, Tim, Stayte, Louise, Vaughan, Alexis, *Sweet and Sour: The Impact of Sugar Production and Consumption on People and the Environment*, Sustain, London, 2000.

Freire, Mdo C., Cannon, G., Sheiham, A., 'An Analysis of the Recommendations on Sugar Consumption between 1961 and 1991', *Rev Saude Publica*, 28(3): 228–37, 1994.

Fried, S.K. and Rao, S.P., 'Sugars, Hypertriglyceridemia, and Cardiovascular Disease', *American Journal of Clinical Nutrition*, 78(4): 873S–880S, 2003.

Freye, Gilberto (Trans.), Samuel Putnam, *The Masters and the Slaves*, Knopf, New York, Originally published 1900, Second Edition 1964.

Galloway, J.H., *The Sugar Cane Industry: An Historical Geography from its Origins to 1914*, CUP, Cambridge, 1989.

Galloway, J.H., 'Botany in the Service of Empire: The Barbados Cane-Breeding Program and the Revival of the Caribbean Sugar Industry, 1880s–1930s', *Annals of the Association of American Geographers*, 86(4): 682–706, 1996.

Galloway, J., 'Sugar', In (Eds.), Kiple, Kenneth and Orvelas Kriemhild, Connee, *The Cambridge World History of Food, Vol I*, pp. 437–49, CUP, Cambridge, 2000.

Gaicottino, Jean-Claude, *Trinidad et Tobago*, Thesis, University of Bordeaux, 1976.

Gibb, H.A.R. (Trans. and Ed.), *Ibn Battuta: Travels in Asia and Africa 1325–1354*, Routledge, London, 1929.

Glasse, Hannah, *The Art of Cookery Made Plain and Easy*, Reprinted with an introduction by Fanny Cradock, Archon Books, SR Publishers, Wakefield, Yorkshire, First published 1747, This edition, 1971.

Glasse, Hannah, *The Compleat Confectioner: or, the Whole Art of Confectionery*, John Exshaw, Dublin, 1760.

Goldsmith, L.A., 'Acute and Subchronic Toxicity of Sucralose', *Food Chem Toxicol*, 38 (Suppl 2): S53–69, 2000.

Grant, Alison, *Bristol and the Sugar Trade*, Longman, Essex, 1981.

Great Barrier Reef Marine Park Authority, *Environmental Status: Water Quality*, Townsville, Queensland, Australia, 23 February 2004. http://www.gbrmpa.gov.au/corp_site/info_services/publications/sotr. water_quality/pressures.html

Greenwood, James, *The Wilds of London*, London, 1876.

Grice, H.C. and Goldsmith, L.A., 'Sucralose – an Overview of the Toxicity Data', *Food Chem Toxicol*, 38 (Suppl 2): S1–6, 2000.

Griffith, Matt, Rice, Tim, Godfrey, Claire, 'Submission to the DEFRA Consultation on Sugar Reform, January 2004', Oxfam, Oxford, 2004.

Grivet, L., D'Hont, A., Dufour, P., Hamon, P., Roques, D., Glaszmann, J.C., 'Comparative Genome Mapping of Sugar Cane with other Species within the Andropogoneae Tribe', *Heredity*, 73: 500–508, 1994.

Grivet, L., D'Hont, A., Roques, D., Feldmann, P., Lanaud, C., Glaszmann, J.C., 'RFLP Mapping in Cultivated Sugarcane (Saccharum spp): Genome Organization in a Highly Polyploid and Aneuploid Interspecific Hybrid', *Genetics*, 142: 987–1000, 1996.

Grivet, L. and Arruda, P., *Sugarcane Genomics: Depicting the Complex Genome of an Important Tropical Crop*, Current opinion in Plant Biology, Sugarcane genomics: 122–7, 2001.

Hall, I.V., *A History of the Sugar Trade in England with Special Attention to the Sugar Trade of Bristol*, Master's Thesis, University of Bristol, 1925.

Handler, Jerome and Lange, Frederick, *Plantation Slavery in Barbados: An Archaeological and Historical Investigation*, Harvard University Press, Harvard, 1999.

Harris, John, 'The Genuine Account of the Dreadful Massacre that befel Captain Codd and his People, in the Ship *Marlborough*', *Felix Farley's Bristol Journal*, 31 March 1753.

Hentzner, P., *A Journey into England*, Strawberry Hill, England, 1757.

Hill, G., *A History of Cyprus*, Vol. 2, CUP, Cambridge, 1948.

Holford, P. and Pfeiffer, C., *Mental Health and Mental Illness – The Nutrition Connection*, ION Press, London, 1996.

Holford, Patrick, *The Optimum Nutrition Bible*, Piatkus, London, 1997.

Holford, Patrick, *The 30 Day Fat Burner Diet: The Revolution in Weight Control*, Piatkus, London, 1999.

Holt, S.H.A., Sandona, N., Brand-Miller, J.C., 'The Effects of Sugar-Free and Sugar-Rich Beverages on Feelings of Fullness and Subsequent Food Intake', *Internation Journal of Food Sciences and Nutrition*, 41: 59–71, 2000.

Hooton, E.A., *Apes, Men, and Morons*, Putnam, New York, 1937.

Hough, L., Phadnis, S.P., 'Enhancement in the Sweetness of Sucrose', *Nature*, 263: 800, 1976.

Howard, Barbara and Wylie-Rosett, Judith, 'Sugar and Cardiovascular Disease: A Statement for Healthcare Professionals from the Committee on the Nutrition of the Council on Nutrition, Physical Activity, and Metabolism of the American Heart Association', *Circulation*, 106: 523–527, 2002.

Hughes, Richard, *A High Wind in Jamaica*, Chatto and Windus, London, 1971.

Humboldt, Alexander von (Trans.), E.C. Otté, Henry G Bohn, *Views of Nature: Or Contemplations on the Sublime Phenomena of Creation; with Scientific Illustrations*, Henry G Bohn, London, 1850.

Ingham, John, 'Now they want to put Jellyfish in Our Sugar', *Daily Express*, 1 October 2000.

Instituto Nacional Do Açúcar, *The Sugar Sector in Mozambique: Current Situation and Future Prospects*, Ministry of Agriculture and Rural Development, Mozambique, 2000.

John, B.A., Wood, S.G., Hawkins, D.R., 'The Pharmacokinetics and Metabolism of Sucralose in the Rabbit', *Food Chem Toxicol*, 38 (Suppl 2): S111–3, 2000.

John, B.A., Wood, S.G., Hawkins, D.R., 'The Pharmacokinetics and Metabolism of Sucralose in the Mouse', *Food Chem Toxicol*, 38 (Suppl 2): S107–10, 2000.

Jones, Donald, *Captain Woodes Rogers' Voyage Round the World 1708–1711*, Bristol Branch of the Historical Association, The University of Bristol, 1992.

Jones, Donald, *Bristol's Sugar Trade and Refining Industry*, Bristol Branch of the Historical Association, The University of Bristol, 1996.

Jones, Tom, *Henry Tate 1819–1899: A Biographical Sketch*, Tate & Lyle, London, 1952.

Kemble, Frances Anne, *Journal of a Residence on a Georgian Plantation in 1838–1839*, (Ed.) John A. Scott, Brown Thrasher Books, The University of Georgia Press, Athens, Georgia 1984.

Kennedy, A.J., 'Genetic Base-broadening in the West Indies Sugar Cane Breeding Programme by the Incorporation of Wild Species', In (Eds.) H.D. Cooper, C. Spillane, T. Hodkin, *Broadening the Genetic Base of Crop*: 283–293 IPGRI/FAO, 2001.

Kille, J.W., Tesh, J.M., McAnulty, P.A., Ross, F.W., Willoughby, C.R., Bailey, G.P., Wilby, O.K., Tesh, S.A., 'Sucralose: Assessment of Teratogenic Potential in the Rat and the Rabbit', *Food Chem Toxicol*, 38 (Suppl 2): S43–52, 2000.

Kille, J.W., Ford, W.C., McAnulty, P., Tesh, J.M., Ross, F.W., Willoughby, C.R., 'Sucralose: Lack of Effects on Sperm Glycolysis and Reproduction in the Rat', *Food Chem Toxicol*, 38 (Suppl 2): S19–29, 2000.

Knight, I., 'The Development and Applications of Sucralose, a New High-Intensity Sweetener', *Can J Physiol Pharmacol*, 72(4): 435–9, 1994.

Leifert, C., 'Naïve, Narrow and Biased . . .', *Guardian*, 24 July 2003.

Lichtenthaler, F.W., Immel, Stefan, Kreis, Uwe, 'Evolution of the Structural Representation of Sucrose', *Starch*, 43(4): S121–132, 1991.

Lichtenthaler, Frieder, 'Emil Fischer's Proof of the Configuration of Sugars: A Centennial Tribute', *Angew Chem Int Ed Engl*, 31: 1541–56, 1992.

Lichtenthaler, Frieder, 'Emil Fischer, His Personality, His Achievements, and His Scientific Progeny', *Eur J Org Chem*: 4095–122, 2002.

Ligon, Richard, *A True and Exact History of Barbadoes*, London, 1657.

Lloyd, David and Wachefeld, David, 'Working to Save Nature's Playground', *Reef Research*, 8(1): 1–6, 1998.

Long, Edward, *History of Jamaica*, London, 1774.

Lu, Y.H., D'Hont, A., Paulet, F., Grivet, L., Arnaud, M., Glaszmann, J.C., 'Molecular Diversity and Genome Structure in Modern Sugarcane Varieties', *Euphytica*, 78: 217–26, 1994.

Ludwig, D.S., Majzoub, J.A., Al-Zahrani, A., Dallal, G.E., Blanco, I., Roberts, S.B., 'High Glycemic Index Foods, Overeating, and Obesity', *Pediatrics*, 103(3): E26, 1999.

Lyle, Colin, *Management Politics and Practice: Sugar Business Experiences*, Unpublished manuscript, 2001.

Lyle, Leonard, Lord of Westbourne, *Mr Cube's Fight against Nationalisation*, Lord Tate & Lyle, London, 1950.

Lyle, Oliver, *The Plaistow Story*, Tate & Lyle, London, 1960.

Marshall, Peter, *Bristol and the Abolition of Slavery: The Politics of Emancipation*, The Bristol Branch of the Historical Association, The University of Bristol, 1975.

Martin, S.I., *Bristol's Slave Trade*, Channel Four Books, London, 1999.

Mathieson, W.L., *British Slavery and its Abolition*, Longmans Green, London, 1926.

May, Robert, *The Art of Cookery*, London, 1654.

Mazumbar, Sucheta, *Sugar and Society in China: Peasants, Technology and the World Market*, Harvard University Press, Cambridge, Mass., 1998.

McDevitt, R.M., Bott, S.J., Harding, M., Coward, W.A., Bluck, L.J., Prentice, A.M., 'De Novo Lipogenesis During Controlled Overfeeding with Sucrose or Glucose in Lean and Obese Women', *American Journal of Clinical Nutrition*, 74(6): 737–46, 2001.

McKendry, M., *Seven Centuries of English Cooking*, Weidenfeld and Nicolson, London, 1973.

Meneight, W.A., *A History of the United Molasses Co. Ltd*, Seel House Press, Liverpool, 1977.

Miller, William, *Journal of the Black Prince 1762–1764*.

Mintz, Sidney, *Sweetness and Power: The Place of Sugar in Modern History*, Penguin, New York, 1986.

Misciagna, G., Centonze, S., Leoci, C., Guerra, V., Cisternino, A.M., Ceo, R., Trevisan, M., 'Diet, Physical Activity, and Gallstones – A Population-Based, Case-Control Study in Southern Italy', *American Journal of Clinical Nutrition*, 69(1): 120–6, 1999.

Mohanty, Priya, Hamouda, Wael, Garg, Rajesh, Aljada, Ahmad, Ghanim, Husam, Dandona, Paresh, 'Glucose Challenge Stimulates Reactive Oxygen Species (ROS) Generation by Leucocytes', *J Clin Endocrinol Metab*, 85: 2970–73, 2000.

Morgan, Kenneth, *Edward Colston and Bristol*, The Bristol Branch of the Historical Association, The University of Bristol, 1999.

Newton, John (Ed.), Bernard Martin, Mark Spurrell, *The Journal of a Slave Trader 1750–1754*, Epworth Press, London, 1962.

Oritz, Fernando, *Cuban Counterpoint*, Knopf, New York, 1947.

Ovadia da Bertinoro, *Viaggio a Gerusalemme*, 1488.

Pares, Richard, *A West India Fortune*, Longmans Green, London, 1950.

Parris, G.K., 'James W. Parris: Discoverer of Sugar Cane Seedlings', *Journal of the Barbados Museum and Historical Society*, XXII: 3–9, 1955.

Parris, James, Letter to the Editor, *The Barbados Liberal*, 8 February 1859.

Peltier, Michael, 'Florida Govenor Bush Signs Contentious Everglades Bill', *Planet Ark*, 16 February 2004.

Pfeiffer, C. and Holford, P., *Mental Illness and Schizophrenia: The Nutrition Connection*, Thorsons, London, 1989.

Polo, Marco, *The Travels of Marco Polo*, (Trans.) Ronald Latham, Penguin, Harmondsworth, this edition 1958.

Pope, Alexander, Letter to Martha Blount, 19 November 1739, In (Ed.) George Sherburn, *The Correspondence of Alexander Pope*, Vol IV 1736–1744, Clarendon Press, Oxford, 1956.

Prentice, A.M., 'Obesity, Dietary Sugars and Physical Activity', In (Ed.) *The Sugar Bureau, Nutrition and Health Aspects of Sugar Consumption: A Case for Reconsideration of the Scientific Basis of Aspects of Current UK Government Advice Relating to Sugar*, appendix three, The Sugar Bureau, London, 1996.

Price, Weston A., *Nutrition and Physical Degeneration: A Comparison of Primitive and Modern Diets and Their Effects*, The American Academy of Applied Nutrition, California, 1948.

Prince, Mary, *The History of Mary Prince: A West Indian Slave Related by Herself*, (Ed.) Moira Ferguson, University of Michigan, Ann Arbor, 1993.

Quick, Allison, Sheiham, Helena, Sheiham, Aubrey, *Sweet Nothings: The Information the public receives about sugar*, 1980.

Raben, A., Vasilaras, T.H., Moller, A.C., Astrup, A., 'Sucrose Compared with Artificial Sweeteners: Different Effects on Ad Libitum Food Intake and Body Weight After 10 wk of Supplementation in Overweight Subjects', *American Journal of Clinical Nutrition*, 76(4): 721–9, 2002.

Raworth, Kate, *The Great EU Sugar Scam: How Europe's Sugar Regime is Devastating Livelihoods in the Developing World*, Oxfam, Oxford, 2002.

Reaven, Gerald, 'Insulin Resistance, Hypertension, and Coronary Heart Disease', *J Clin Hypertens*, 5(4): 269–74, 2003.

Reaven, G.M., 'Role of Insulin Resistance in Human Disease', *Diabetes*, 37(12): 1595–607, 1988.

Reaven, G., Abbasi, F., McLaughlin, T., 'Obesity, Insulin Resistance, and Cardiovascular Disease', *Recent Prog Horm Res*, 59: 207–23, 2004.

Reddy, M. and Yanagida, J., 'Welfare Implications of Alternative Markets: A Case Study of Fiji's Sugar Industry', *Journal of South Pacific Agriculture*, 15(1): 61–8, 1998.

Reef Management News, 'Authority Supports "Greener" Sugar Industry', *Reef Research*, 8(2): 1–3, 1998.

Revill, Jo and Ahmed, Kamal, 'The Junk Food Timebomb that Threatens a New Generation', *Observer*, 9 November 2003.

Revill, Jo and Harris, Paul, 'America Stirs Up a Sugar Rebellion', *Observer*, 18 January 2004.

Revill, Jo and Harris, Paul, 'US Sugar Barons "Block Global War on Obesity"', *Observer*, 18 January 2004.

Richardson, David, *The Bristol Slave Traders: A Collective Portrait*, Bristol Branch of the Historical Association, The University of Bristol, 1985.

Roach, T., 'Sugar canes', In (Eds.) Smartt, Joseph, Simmonds, Norman, *The Evolution of Crop Plants* (Second Edition), Longman, London, 1995.

Roberts, A., Renwick, A.G., Sims, J., Snodin, D.J., 'Sucralose Metabolism and Pharmacokinetics in Man', *Food Chemical Toxicology*, 38 (Suppl 2): S31–S41, 2000.

Rohwer, Johann M., Botha, Frederik C., 'Analysis of Sucrose Accumulation in the Sugar Cane Culm on the Basis of in vitro Kinetic Data', *Biochemical Journal*, 358: 437–45, 2001.

Sandiford, Keith (Ed.), *The Cultural Politics of Sugar: Caribbean Slavery and Narratives of Colonialism*, CUP, Cambridge, 1947, reprinted 2000.

Saris, W.H., 'Sugars, Energy Metabolism, and Body Weight Control', *American Journal of Clinical Nutrition*, 78(4): 850S–857S, 2003.

Saris, W.H., 'Glycemic Carbohydrate and Body Weight Regulation', *Nutr Rev*, 61(5 Pt 2): S10–6, 2003.

Schenk, P.M., Remans, T., Sagi, L., Elliott, A.R., Dietzgen, R.G., Swennen, R., Ebert, P.R., Grof, C.P., Manners, J.M., 'Promoters for pregenomic RNA of banana streak badnavirus are active for transgene expression in monocot and dicot plants', *Plant Mol Biol*, 47(3): 399–412, 2001.

Shauss, A.G., 'Nutrition and Behavior', *Journal of Applied Nutrition*, 35(1): 30–5, 1983, and *MIT Conference Proceedings on Research Strategies for Assessing the Behavioural Effects of Foods and Nutrients*, 1982.

Shreuder, Yda, 'The Influence of the Dutch Colonial Trade on Barbados in the Seventeenth Century', *Journal of the Barbados Museum and Historical Society*, XLVIII: 43–63, 2002.

Sloane, Hans, *A Voyage to the Islands of Madera, Barbados, Nieves, St Christophers and Jamaica*, Vol I, London, 1701.

Sims, J., Roberts, A., Daniel, J.W., Renwick, A.G., 'The Metabolic Fate of Sucralose in Rats', *Food Chem Toxicol*, 38 (Suppl 2): S115–21, 2000.

Smith, Adam, *An Enquiry into the Nature and Causes of the Wealth of Nations*, Edinburgh, 1853.

Smith, Jeremy, 'Sweet Smell of Excess', *Ecologist*, November 2003.

Smith, Mary, *Baba of Karo: A Woman of the Moslem Hausa*, Faber, London, 1954.

Stevenson, G.C., 'Sugar Cane Varieties in Barbados: An Historical Review', *Journal of the Barbados Museum and Historical Society*, XXVI: 67–102, 1959.

Stevenson, G.C., *Genetics and Breeding of Sugar Cane*, Longmans Green, London, 1965.

Strong, L.A.G., *The Story of Sugar*, Weidenfeld and Nicolson, London, 1954.

Taviani, Emilio Paolo, *Christopher Columbus: The Grand Design*, Orbis Publishing Ltd, London, 1974.

Twain, Mark, *Life on the Mississippi*, 1883.

Tussac, F., *Flora Antillarum*, Paris, 1800.

Tyron, Thomas, *Friendly Advice to the Gentlemen Planters of the West Indies*, Philotheos Phisiologus, London, 1684.

United States Beet Sugar Association, *The Beet Sugar Story*, United States Beet Sugar Association, 1959.

Vaughan, Alexis, 'Sugar, Trade and Europe: A Discussion Paper on the Impact of European Sugar Policies on Poor Countries', Sustain, London, 2000.

Venner, Tobias, *Via Recta ad Vitam Longam*, London, 1620.

Vermunt, S.H., Pasman, W.J., Schaafsma, G., Kardinaal, A.F., 'Effects of Sugar Intake on Body Weight: A review', *Obes Rev*, 4(2): 91–9, 2003.

Vettore, A.L., da Silva, F.R., Kemper, E.L., Arruda, P., 'The Libraries that made SUCEST', *Genetics and Molecular Biology*, 24(1–4): 1–7, 2001.

Warner, Oliver, *William Wilberforce and his Times*, BT Batsford Ltd, London, 1962.

Watkins, Kevin, *Dumping on the World: How EU Sugar Policies Hurt Poor Countries*, Oxfam International, Oxford, 2004.

Watson, Andrew, *Agricultural Innovation in the Early Islamic World*, CUP, Cambridge, 1983.

Watson, J.A., *A Hundred Years of Sugar Refining: The Story of Love Lane Refinery 1872–1972*, Tate & Lyle Refineries Ltd, Liverpool, 1973.

Watson, J.A., *Talk of Many Things: Random Notes concerning Henry Tate and Love Lane*, Tate & Lyle Refineries Ltd, Liverpool, 1985.

Wesley, John, *Thoughts Upon Slavery*, 1774.

Willett, Walter, Stampfer, Meir J., 'Rebuilding the Food Pyramid', *Scientific American*, January 2003.

Williams, E., *Capitalism and Slavery*, University of North Carolina Press Chapel Hill, NC, 1944.

Wood, S.G., John, B.A., Hawkins, D.R., 'The Pharmacokinetics and Metabolism of Sucralose in the Dog', *Food Chem Toxicol*, 38 (Suppl 2): S99–106, 2000.

WHO, *Obesity: Preventing and Managing the Global Epidemic*, WHO, Geneva, 2004.

Wray, Leonard, *The Practical Sugar Planter*, Smith, Elder and Co, London, 1848.

Young, Filson, *Christopher Columbus and the New World of his Discovery: A Narrative*, Vol. I, E. Grant Richards, London, 1906.

Yudkin, John, Edelman, Jack, Hough, Leslie, (Eds.), *Sugar: Chemical, Biological and Nutritional Aspects of Sucrose*, Butterworths, London, 1971.

Yudkin, John, *Pure, White and Deadly: The Problem of Sugar*, Davis-Poynter, London, 1972.

Wu Zimu, *The Past Seems a Dream*, 1275.

INDEX